Automation and Robotics in the Architecture, Engineering, and Construction Industry

Houtan Jebelli
Mahmoud Habibnezhad
Shayan Shayesteh • Somayeh Asadi
SangHyun Lee
Editors

Automation and Robotics in the Architecture, Engineering, and Construction Industry

Springer

Editors
Houtan Jebelli
Department of Architectural Engineering
Pennsylvania State University
University Park, PA, USA

Mahmoud Habibnezhad
Department of Architectural Engineering
Pennsylvania State University
University Park, PA, USA

Shayan Shayesteh
Department of Architectural Engineering
Pennsylvania State University
University Park, PA, USA

Somayeh Asadi
Department of Architectural Engineering
Pennsylvania State University
University Park, PA, USA

SangHyun Lee
Civil and Environmental Engineering
University of Michigan–Ann Arbor
Ann Arbor, MI, USA

ISBN 978-3-030-77162-1 ISBN 978-3-030-77163-8 (eBook)
https://doi.org/10.1007/978-3-030-77163-8

This Springer imprint is published by the registered company Springer Nature Switzerland AG
The registered company address is: Gewerbestrasse 11, 6330 Cham, Switzerland

Preface

In the early twentieth century, construction witnessed the emergence of heavy machines at job sites, a revolutionary transformation from animal and human-powered machines to those powered by diesel engines, hydraulic pumps, and electric motors. A century later, yet another transformation began, this time directed at adopting intelligent machines, autonomous vehicles, and perceptive cyber-physical systems. This paradigmatic shift is intended to respond to stagnant productivity, labor shortage, aging workforce, and unsafe environments of construction sites, which has become possible due to the advent of sensing technologies, computer processing powers, and artificial intelligence. In the last few years, the number of start-up companies in robotics has multiplied, particularly in autonomous mobility. Even the giants of construction machinery have started to make investments in emergent technology and intelligent machines. Caterpillar Inc. has recently bought a robotics-based company, Marble Robot, to boost its research efforts on the next generation of unmanned heavy machines. Bobcat Company has teamed up with QinetiQ North America to transform compact equipment into a remote-control vehicle. Volvo Group has invested in Apex.AI to become one of the forefronts of safety-certified autonomous systems. The exponential growth of robotic systems is not limited to unmanned vehicles. 3D printing robots have already found their way into the construction industry, with the capacity to build houses in a few days with almost half the price of the comparable ones. Construction robotics has designed and engineered a commercially available brick-laying robotic arm capable of laying bricks with lower cost and labor. The same company has introduced a lift-assist robot for lifting and placing heavy materials with minimal effort. These advancements in construction robotics have provided an unparallel opportunity to automate repetitive tasks with robots, handle strenuous activities using wearable and lift-assist machines, replace ambulatory surveys and inspections with unmanned aerial vehicles–based monitoring, and perform dangerous tasks remotely through telerobotics.

As promising as these innovations appear to be, the scalability, extendibility, and ubiquity of robotics systems are still unknown. Several ambiguities concerning the role of the human workforce in future job sites need to be resolved, which merits immediate attention and investigation. Suppose construction robots become prevalent in the fields. In that case, they will have sweeping effects on the physiological and psychophysiological states of human workers, safety and productivity expectations, and human-robot trusts as well as the necessary competencies of workers, engineers, and other construction personnel. Multiple questions need to be addressed promptly. What are the societal, economic, and environmental impacts of these systems? How do workers perceive robots at the job sites and vice versa? What education and training should be developed, and how they need to be conveyed? What are the directions of improvements of construction robots, and how the related challenges can be timely and effectively addressed? What areas of construction can most benefit from the adoption of robotics systems? These fundamental questions help the industry prepare for introducing, implementing, and continuing robotics solutions in construction working environments.

Moreover, the ever-existing safety challenges in the construction industry may manifest themselves in new forms in construction robotics, with a myriad of uncertainties and ambiguities. In the last 20 years, reports have unequivocally demonstrated that the construction industry has not been able to reduce the number of death and injuries at construction sites. Specifically, within the same time frame, the statistics show that the fatality rates of fieldworkers have generally remained constant, somehow invariable to the safety practices, regulations, and protocols in place. This immutable safety condition has been mainly linked to the dynamic and changing environment of construction sites wherein construction workers perform an array of manual activities in isolation from or collaboration with heavy machinery. While construction robotics aims to address these challenges, they require human supervision/intervention to deliver the work. The reason lies in the design complexities of robots in working or even surviving in the dynamically changing environment of construction sites. As a result, human-robot interaction/collaboration is required. Yet, as history teaches, these interactions can lead to accidents, similar to those that occur every day at construction job sites. Aggravatedly, the control of heavy machinery is by human operators, while the movement of robots is based on predefined codes with possible model- and system-level errors. This can have a detrimental effect on human-robot collaboration safety.

Moreover, accidents can happen due to the soar of cognitive workload and mental and physiological stress during interaction with robots, possibly the direct effects of the unparallel performance of robots compared to their human peers as well as worker-robot mistrust and misadaptation. Furthermore, human workers cannot recognize or predict the robot's motion legibility because of their expectations of the robot's human-like movement, which, of course, deviates from reality. These remarks stress the need for comprehensive investigations to understand influential factors affecting safety in human-robot interactions, which has not been most likely contemplated at any depth by the early adopters of construction robots.

Based on these gaps, the editors of this book were persuaded that there is a need for comprehensive, evidence-based insight and view of current and future automation and robotics in the architecture, engineering, and construction industry. As such, the editors compiled this book to illuminate the challenges, opportunities, and needs in adopting robots in the construction industry. This book attempts to construct the road map and delineate directions of improvements for a successful transition towards construction robotization. It is the fruition of several scientific investigations on priorities and barriers towards robotic technologies and deployment in construction. Notably, the collection leverages the research findings of several contributors from the Pennsylvania State University, University of Michigan, University of Florida, University of Alabama, University of Illinois at Chicago, Wentworth Institute of Technology, Washington State University, and Institute of Construction Materials.

Even though the contents and orders of the chapters follow a logical flow, the book does not have a textbook structure. This allows the readers to jump to the preferred chapter and understand the deliverables without much hassle. The presented knowledge in each chapter is supported by adequate explanation and elaboration, making it easier for the layperson to understand. But, there are very few chapters with algorithms and mathematical modeling that might be overwhelming for ordinary readers. In those, the authors attempted to provide the importance of research, the gaps in the existing methods, references to other work, in-depth discussion, and thought-provoking conclusion, all to facilitate the comprehension of the presented science.

We hope this book can deliver its message to a wide group of industry professionals and academic scholars and anyone interested in obtaining a better understanding of construction robotics and the potential opportunities and the daunting challenges. By reading this book, learners can have a new insight into construction robots and workplaces of tomorrow, which can help them make better decisions on the required competencies for future jobs in construction. Similarly, educators and practitioners can leverage the information presented in this compilation and appropriately infuse them into curriculums and training at different scales. Finally, it could be of interest for field engineers who had come across smart machines, construction robots, and aerial and terrestrial vehicles in the field but not studied them beforehand.

University Park, PA, USA

Houtan Jebelli
Mahmoud Habibnezhad
Shayan Shayesteh
Somayeh Asadi

Ann Arbor, MI, USA

SangHyun Lee

Contents

1 **Automation and Robotics Technologies Deployment Trends in Construction** 1
Fopefoluwa M. Bademosi and Raja R. A. Issa

2 **Essential Knowledge, Skills, and Abilities Required for Talent Cultivation in Construction Automation and Robotics** 31
Fopefoluwa M. Bademosi and Raja R. A. Issa

3 **Robotics and Automation in Construction (RAC): Priorities and Barriers Toward Productivity Improvement in Civil Infrastructure Projects** 59
Abolfazl Seyrfar, Hossein Ataei, and Ibrahim Osman

4 **Construction Automation and Sustainable Development** 73
Mohammadsoroush Tafazzoli

5 **A Briefing on Technological, Societal, and Economical Aspects of Automation in Construction** 97
Farid Sartipi

6 **Human-Drone Interaction (HDI): Opportunities and Considerations in Construction** 111
Gilles Albeaino, Masoud Gheisari, and Raja R. A. Issa

7 **Safety Challenges of Human-Drone Interactions on Construction Jobsites** 143
Idris Jeelani and Masoud Gheisari

8 **Assessing the Role of Wearable Robotics in the Construction Industry: Potential Safety Benefits, Opportunities, and Implementation Barriers** 165
Ifeanyi Okpala, Chukwuma Nnaji, Omobolanle Ogunseiju, and Abiola Akanmu

9 Worker-Aware Task Planning for Construction Robots: A Physiologically Based Communication Channel Interface 181
Yizhi Liu, Mahmoud Habibnezhad, and Houtan Jebelli

10 Workers' Trust in Collaborative Construction Robots: EEG-Based Trust Recognition in an Immersive Environment 201
Shayan Shayesteh, Amit Ojha, and Houtan Jebelli

Index 217

Chapter 1
Automation and Robotics Technologies Deployment Trends in Construction

Fopefoluwa M. Bademosi and Raja R. A. Issa

Abstract The construction industry has remained very much less digitized than similar mature industries, with a good number of its activities being laborious and monotonous. With the construction industry gradually entering the era of industrialization and digitization, there is an opportunity to increasingly change the traditional approach of designing, constructing, operating, and maintaining buildings and infrastructure. The era of technological disruption ushers in the penetration of innovative construction tools, including automation and robotics technologies, in the industry. In construction, automation and robotics processes involve the autonomous execution of tasks and processes through intelligent programming and control of machines. This chapter explores the current state of automation and robotics technologies deployment in the construction industry and discusses potential automation and robotics applications in different aspects of construction projects. Construction companies can use these technologies to benchmark their technology adoption efforts.

Keywords Construction · Automation · Robotics · Industrialization · Digitization · Industry 4.0

F. M. Bademosi (✉)
Wentworth Institute of Technology, Boston, MA, USA
e-mail: bademosif@wit.edu

R. R. A. Issa
University of Florida, Gainesville, FL, USA
e-mail: raymond-issa@ufl.edu

H. Jebelli et al. (eds.), *Automation and Robotics in the Architecture, Engineering, and Construction Industry*, https://doi.org/10.1007/978-3-030-77163-8_1

1.1 Introduction

The construction industry makes up a substantial fraction of the economic sector across continents, with the construction industry's global expenditure estimated at $11.4 trillion and expected to approach $14 trillion in 2025 [1]. However, unlike other mature industries, the construction industry still operates around numerous arduous and tedious tasks, which has led to declining productivity levels and sluggish growth. To compete effectively in the wake of rising urbanization and globalization, escalating market competition, and technology developments, the industry calls for efficacious construction organizations, practices, and advanced techniques [2].

Today, the world has a population of about 7.4 billion, with 55% of the populace residing in urban regions. Forecasts predict that urbanization, the progressive redistribution of the population from rural to urban regions, and the world's global population's growth could increase the urban population's proportion to 68% by 2050 [3]. Cities are anticipated to grow at a rate of 200,000 people daily, and the architecture, engineering, construction, and operations (AECO) industry is expected to build an incremental 1000 buildings a day and invest 3.7 trillion US dollars in infrastructure until 2035 to address the needs of the projected growth [2, 4]. This need for buildings and infrastructure will create an excellent opportunity for the AECO industry to explore further industrialization and digitization.

One of the industry's predominant challenges is declining productivity levels [1]. The deployment of cutting-edge equipment and systems is crucial for the construction industry, as an upturn in overall productivity levels will substantially impact the economy at a macro level. A potential avenue for groundbreaking technologies in the industry is the application of automation and robotics technologies to enhance efficiency, output, health and safety, and quality factors. Despite these technologies' potential to improve construction processes, the industry has been slower in adopting and adapting to new technologies than other mature industries. One of the contributing factors to the slow adoption is the limited research studies that have endeavored to uncover the shift and trends in automation and robotics technologies.

This research builds on existing literature and addresses previous research studies' limitations by establishing an understanding of automation and robotics technologies as applicable to construction. The deployment of these technologies in the existing industrialization and digitization endeavors within the industry is also assessed. As these technologies continue to evolve significantly, it is becoming critical to understand their impact on existing work practices. Recent advancements in automation and robotics technologies were identified through literature review and their applications in the construction industry, especially related to construction process automation and efficiency and productivity enhancements. Automation and robotics technologies are expected to keep evolving and will require continuous evaluation. However, this research lays the foundation for promoting the integration of automation and robotics technologies in construction.

The chapter is organized as follows. The remainder of Sect. 1.1 highlights the major problems facing the construction industry, the eras of construction disruption, and the anticipated nature of the future of work in the construction industry. Key definitions of automation and robotics concepts and how they are applied in the construction industry and the different categories of automation and robotics technologies are presented in Sect. 1.2. Section 1.3 contains the results from semi-structured interviews conducted with construction technology and innovation specialists to determine where the identified automation and robotics technologies stand in current industrialization and digitization endeavors in the industry. The different applications of automation and robotics technologies in construction are explored in Sect. 1.4. Finally, conclusions are provided in Sect. 1.5.

1.1.1 Major Problems Facing the Construction Industry

Productivity in the construction industry has been steadily declining over the past decades, especially compared to the manufacturing industry, as seen in Fig. 1.1 [5]. An improvement in productivity levels in construction can be achieved through the implementation of advanced technologies and processes. Other challenges that conventional construction methods have not yet been able to solve include decreased quality of the overall products, skilled labor shortages, unfavorable working conditions, safety concerns, defect rates, cost overruns, an aging workforce, and copious

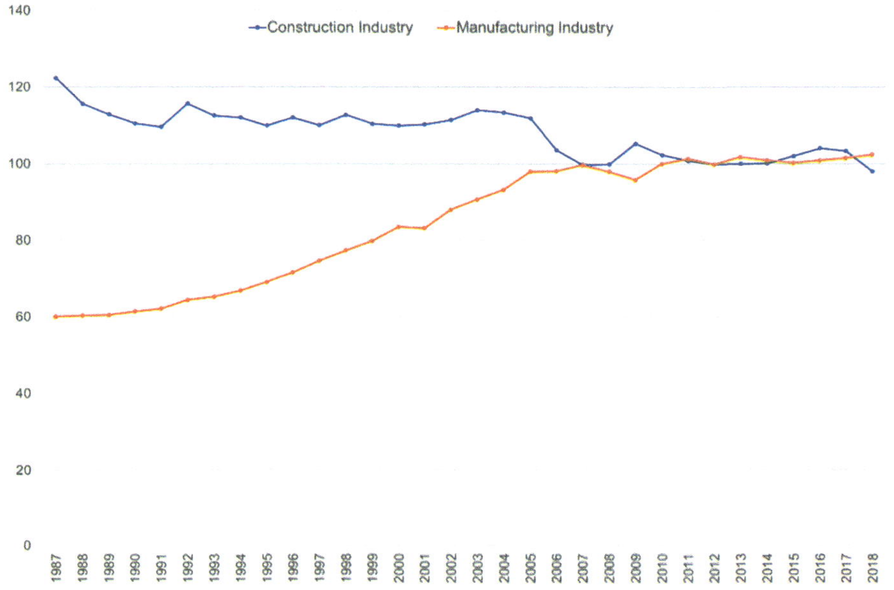

Fig. 1.1 Construction and manufacturing industry productivity trends. (Adapted from [5])

depletion of raw elements, presenting an opportunity for innovative solutions within the industry [6–8].

For the construction industry to profitably compete and remain significant under the strain of increasing urbanization, globalization, industry opposition, and technological innovations in the era of industrialization, more innovative, practical, and effective building processes are necessary [2]. The construction industry is an important economic sector, and the industry needs to place value on efficiency to attain and sustain productivity and economic growth. Despite the potentials innovative technologies have in resolving the industry's problems, the construction industry has adopted new technologies much slower than other industries [9]. One of the reasons is that the construction industry has low capital investment margins and capital intensity [10]. New technologies are also considered inferior in performance at the initial phase of implementation due to technical, regulatory, and economic barriers in conjunction with their minimal assimilation into an economic climate dominated by seasoned and traditional technologies.

The aforementioned problems infer that traditional construction has reached a limit in technological performance in the industry. Automation and robotics technologies have the potential to ameliorate many aspects of the industry. The potential capacity of delivering better quality products at lower unit costs and an increased rate could inevitably improve the competitiveness of the industry on a global scale. With automation and robotics, jobsites could be accommodated in a safer environment, resulting in a higher level of control over the construction process, greater performance efficiency, and more consistent outcomes [7].

The study of automation and robotics technologies in construction has become an essential research area within the AECO industry. Since the dawn of computer-aided design (CAD), followed by building information modeling (BIM), the AECO industry has undergone profound changes in the extent of technological integration. These technological shifts have resulted in favorable changes in the industry, enabling the utilization of innovative construction technologies to improve the efficiency and productivity levels of construction processes and products.

1.1.2 Eras of Construction Disruption

The AECO industry has encountered a rise in the number of modern technologies available for deployment. The development of innovative technological processes in the industry has progressed, and their impact on the industry in its entirety has escalated to the delivery of an improved built environment. From the advent of CAD through to BIM, cloud computing, and generative design, construction disruption can be categorized into four eras: documentation, optimization, and connection, with the future being the industrialization era, as shown in Fig. 1.2 [11].

CAD software has had a notable significance in the AECO industry following the emergence and prevalence of computing in education in the 1950s and the influx of personal computing in the 1980s [12]. Although these tools supported designers in

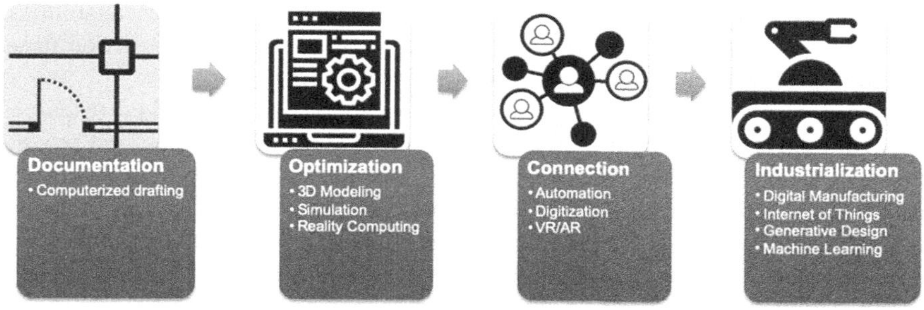

Fig. 1.2 Eras of construction disruption. (Adapted from [11])

producing technical drawings faster, the traditional design process did not radically change. The introduction of BIM, the practice of creating a computer-generated representation of a structure with precise geometrical patterns and pertinent design and construction data [13], in the 1990s, led to the "Era of Optimization." Conceptual BIM began in the late twentieth century but did not appear in a tangible form until the twenty-first century.

In the last few years, the advent of new-age technologies and processes has enabled transitioning the industry from modeling to information sharing, enabling project stakeholders to collaborate more strongly. This practice has been termed *Connected BIM*, the interconnection of all construction processes, team members, and technologies in the information environment [14]. With *Connected BIM*, smart and efficient management of the entire project lifecycle is possible, significantly improving project outcomes [14].

Over the past decade, several mature industries have encountered advancements in digital technology, resulting in notable transformations. This evolution led to the latest technological era, recognized as the Fourth Industrial Revolution (Industry 4.0), otherwise called the era of industrialization [15]. Recently, advanced construction technologies have started to penetrate the industry, progressively transforming how buildings and infrastructure are designed, constructed, operated, and maintained [16].

1.1.3 The Future of Work

The technologies ascribed to the era of industrialization and digitization are vital in solving the low productivity levels and inferior quality of products in the construction industry [7]. Construction-related businesses are progressively integrating innovative technologies and systems and the latest software unified with the focal hub of connected BIM on projects and jobsites. The era of industrialization and digitization offers many advantages to the industry, the ecosystem, and the global market [16]. However, the widespread implementation of automation and robotics technologies and processes in the industry is restricted, despite their capabilities. The implementation of innovative technologies is essential for construction firms that aim to attain economic and competitive advantage.

Firms must assess the feasibilities of advanced technologies and unique resources and aggressively transform the internal strategies, methods, protocols, and organization model of the company to succeed in the era of industrialization appropriately, driven by innovative technologies. Companies must act in order to remain relevant in this new era of industrialization. Some essential calls to action include attracting new talented experts and developing necessary skills and abilities, collaborating and consolidating efforts across the value chain of the construction industry, promoting widespread adoption of advanced technologies, and increasing the end-to-end applications of information and digital models in construction operations [2].

1.2 Automation and Robotics Technologies

To attain a clear understanding of automation and robotics concepts and how they are applied in the construction industry, it is imperative to determine a succinct and satisfactory definition of the terms. Firstly, broad definitions of automation and robotics are presented, followed by specific definitions correlating them to practical applications in construction. The summary of the definitions of the terms is then presented from the review of available literature, from which the definition of construction automation and robotics for this research was coined.

A robot is defined as a machine that moves independently and executes complex tasks typically assigned to humans [17]. It is also defined as an automatic device that carries out complicated tasks that are often repetitive. A robot is usually programmed via automatic controls that enable it to imitate the human range of motions, equipped with multi-functional handlers explicitly intended for moving materials, instruments, or parts required to complete tasks. Robotics has been defined as the science of designing, manufacturing, and operating robots computed intelligently to complete tedious tasks, simulated via automated controls [18]. In several industries, robots are employed to execute jobs considered dangerous and monotonous, and as these technologies advance, robots are expected to be used to complete more complex job tasks.

Another definition of robotics is the domain that permits the fabrication of robots [19]. Robots are typically devised to conduct several related duties instead of humans. Therefore, they should be constructed with the perception and behavior that enables them to perform work effectively. Furthermore, robotics has been defined as the science of incorporating the education background, intelligence, and creative thinking of computer, manufacturing, mechanical, industrial, and electrical engineering to design, build, and operate robots to carry out work [20, 21].

Automation has been defined historically as the application of automated processes and machines, principally for production or data-processing operations that demand limited or no human interference during routine procedures [22].

Automation is also defined as a process of implementing autonomous devices to execute production-related operations automatically through a feedback process in which sensors monitor the operations of a system and transmit indicators to a computerized device that instinctively controls the operation [23].

Furthermore, automation is defined as the performance of an apparatus, process, or system operated by an automatically controlled mechanical or electronic device designed to supplant human labor [24]. A concise and acceptable definition of automation is an automatic method of performing a sequence of assigned tasks with programmable devices [25]. Although integrating machines into a production process is referred to as mechanization, automation goes a step further to integrate machines whose behaviors can be monitored and controlled by a computer-assisted program [26].

Construction is defined as the means of constructing or the industry where buildings are erected [27]; the act or manner of constructing; the exploit of building or erecting a structure; the act of devising, forming, and fabricating; the composition, method, and style of building; or assembling the parts of a thing [28]. Construction automation is the application of automated and computerized techniques in construction to perform mechanical operations aimed at reducing time, labor intensity, and risk of exposure to harm while preserving and even enhancing the quality of the finished products [19]. Construction robots are intelligent machines operated by smart controls but range in levels of sophistication. However, they are generally intended to enhance speed and boost the precision of construction processes [29].

Both the terms automation and robotics have been broadly recognized in the construction industry and customarily describe mechanization, unmanned processes, and robotization of construction tasks and processes [7]. Therefore, the definition of construction automation and robotics can be synthesized from the definitions above and summarized simply as the application of autonomous machines that carry out construction tasks and operations automatically through intelligent programming and controls.

Automation and robotics technologies have been implemented as a strategic solution to productivity challenges as early as the 1960s [30]. However, research in the field of automation and robotics in construction has advanced significantly since then. In recent years, there has been a rise in the innovative technologies deployed in automation and robotics. While some of these technologies have been widely utilized in construction, many are up and coming and just surfacing in the market [16]. Moreover, the manufacturing industry has pioneered many of the automation and robotics technologies being implemented in the construction industry [31]. In the last few years, the expression "Industry 4.0" has been applied as a model to illustrate the latest advancements in manufacturing. Researchers [16] suggested applying the Industry 4.0 concept as a point of reference to categorize the innovative technologies guiding the era of industrialization and digitization in construction.

Several researchers [16, 32] have conducted studies that gave a valid account of the Industry 4.0 technologies in construction by completing a content-based assessment of existing literature related to the topic. The results encompassed the prominent cutting-edge subjects in automation and robotics. The identified automation and robotics technologies were further grouped into four categories: realty computing technologies, visualization technologies, automation technologies, and intelligent systems. The technologies have different levels of automation (LOA), ranging from teleoperation to full autonomy. These LOA influence the way in which humans and automation and robotics technologies interact with one another [33].

1.2.1 Reality Computing Technologies

1.2.1.1 Unmanned Aerial Systems (UASs)

The application of UASs in the construction industry has unexplored potential, offering researchers and industry professionals a new tool. The integration of UASs into AECO industry workflows has been a popular area of research interest in recent years, as industry professionals and researchers continue to demonstrate and develop additional use cases for UASs. The uses for UASs have ranged from site documentation to building inspections and even disaster site assessment [34]. When outfitted with high-resolution cameras and sensors, UASs can be used to survey sites before construction commences, monitor the progress of the project, and monitor inventory and stockpile. Other applications include transportation, logistics, health and safety assessment, and the location of possible hazards on the jobsite [34, 35]. Implementation of UASs on construction sites can decrease the exposure of workers to dangers associated with the use of scaffolds and other aerial work platforms [35].

1.2.1.2 Laser Scanners

The technology prevalent in capturing real-world data and conditions subsequently converted for use in a digital environment in the AECO industry is laser scanning. Laser scanning enables the user to capture three-dimensional point cloud data to develop an as-built representation of a space, building, or object [36]. Several industries have demonstrated potential use cases for laser scanning, including process plants, building renovation, transportation, offshore oil production facility development, and utilities [37]. There is an extensive range of applications for laser scanning in the AECO industry. The multiple applications of laser scanning for AECO projects can be categorized into four: three-dimensional modeling and mapping at an accelerated urban scale, management of infrastructure resources, monitoring of construction site operations, and inspection and analysis of structural elements [38].

1.2.1.3 Rovers

Rovers are autonomous machines equipped with high-definition cameras and reality capture sensors that allow them to navigate construction sites [39]. A typical characteristic of construction sites is the extensive gross floor area, making job tasks more complicated and challenging to manage. The vast expanse often results in much time spent completing daily tasks such as site documentation, progress monitoring, and inspections. These processes can be improved efficiently by autonomously employing rovers for quality control purposes, to capture photographs of the active surroundings on the jobsites, validate a situation on-site with BIM data, and identify problems on-site for further site investigation [40]. An emerging leader in this field is the Doxel AI, which uses rovers to comb through construction sites to capture images and laser scans daily and gain the most accurate, up-to-date information. The algorithm of this construction productivity solution measures construction progress, inspects the quality of work, and identifies issues promptly [41].

1.2.1.4 360-Degree Technology

360-Degree technology is the concept of working with 360-degree panoramas captured using 360-degree cameras. 360-Degree panoramas produce an uninterrupted view of the real-world environment resulting in an outcome that is indistinguishable from actuality, providing the user with a sense of presence [42]. 360-Degree panoramas have been employed for several construction applications, including construction progress documentation [43], interactive learning on construction sites [44], reality backdrop for information augmentation [45], safety hazards identification and visualization on construction sites [42], and virtual jobsite visits [46].

1.2.2 Visualization Technologies

1.2.2.1 Virtual Reality (VR)

VR is a three-dimensional environment that permits construction participants to walk around a building model and relate with its components in a computer-generated setting [47]. VR fosters a constructive atmosphere for identifying and comprehending the spatiotemporal constraints inherent in construction processes and for picturing and exploring the three-dimensional layout of construction assemblies. Through a VR headset, the participants can circumnavigate the project by walking from place to place, ascending and descending stairs, and even hovering around for an overview of the virtual model. This virtual experience allows the user to get acquainted with the BIM model and interact with the model and control some of the model attributes [48]. The applications of VR headsets in construction include cost estimating [49], site layout optimization [50], progress monitoring [51], and historic building preservation [52]. Furthermore, VR can be used in AECO education and training, such as developing a virtual learning environment that is entirely risk-free for conducting construction safety training [53].

1.2.2.2 Augmented Reality (AR)

AR is an advancing technology that has achieved considerable importance in research and practice in the AECO industry over time. Fundamentally, AR is regarded as a channel between VR and telepresence as it magnifies realism rather than substitute it [54, 55]. AR can be defined as the amplification of the real environment with data from the virtual environment, that is, augmenting reality with virtuality [56]. AR can be used to detect construction defects when combined with BIM during defect management [57, 58] and real-time collaboration and interactions on the jobsite [59]. Applications of AR in construction have advanced with the introduction of Google Glasses in 2014 and Microsoft HoloLens in 2016 [60]. Furthermore, AR has been recognized as an innovative instruction tool in education and training to enhance understanding of construction scenarios, tasks, and processes [61].

1.2.2.3 Mixed Reality (MR)

Advances in the interaction with construction project data can be accomplished via MR, which is the more extensive application of VR-related technologies. MR indicates the reality-virtuality continuum where virtual contents may be combined with the view of a user in the real environment scene in different proportions [62, 63]. Moreover, it allows for added prospects regarding the interactions of the project members with the virtual model throughout the lifecycle of the project. With the use of intuitive human-computer devices and a successful amalgamation of the virtual world with the real world, MR systems can be applied to improve the decision-making process during the design, planning, execution, monitoring, and inspection phases of construction [62].

1.2.2.4 Digital Twins (DT)

DT, which are precise, duplicate models of physical objects, machines, environments, or processes, are transforming the construction industry [64]. The sophisticated virtual models are driven by real-time data collected from sensors to mirror almost every aspect of a physical object of interest to foresee issues, monitor performance, and increase efficiency [65]. DT can be used to evaluate the current state of a digital duplicate of a physical object or environment and to predict multiple future outcomes [64]. Therefore, the data used for a DT needs to be an accurate representation of the static physical features and the dynamic states of the object or environment. Some of the applications of DT in construction include lifecycle management, real-time construction monitoring, optimized decision-making, performance prediction, data capturing and integration, and optimized quality and product reliability [66]. Furthermore, investing in DT is expected to increase productivity due to predictive analytics and value-added services [65].

1.2.3 Automation Technology

1.2.3.1 Robotic Systems

Generally, robots are focused on performing a narrow selection of repetitive construction activities, predominantly material handling. However, the emerging technologies used in construction include roadwork robots, forklift robots, bricklaying robots, 3D printing robots, soldering robots, aerial robots, demolition robots, robotic total stations, and exoskeletons [21]. These are called single-task construction robots (STCRs) [30]. Construction robotics developed two popular construction robotic systems, a material unit lift enhancer (MULE), Mule135, intended for handling and placing materials weighing up to 135 pounds on jobsites, and a semi-automated mason (SAM), SAM100, a bricklaying robot intended and engineered for onsite masonry construction [30]. Also, a selection of sensors, including cameras and laser scanners, can be wired to robots, which allows for the collection of data on project operations such as progress monitoring, safety monitoring, surveying, inspection, and maintenance. Subsequently, the collected data can be supplied to the control unit and examined to modify the operation of the machine to achieve greater performance and precision [20].

1.2.3.2 Additive Manufacturing (AM)

Additive manufacturing (AM), also called 3D printing, is at the forefront of digital construction. It involves the production of objects by creating structures from small layers of material. AM offers the possibility to translate the 3D model of a building element directly into a completed product by making use of a broad selection of resources such as concrete, polymer, ceramic, glass, steel, and several others [67]. Without the constraints of standard building elements, contractors can tackle project challenges by providing optimum resolutions that require minimum effort and zero waste [67].

1.2.3.3 Prefabrication and Modularization

Prefabrication is the assembly of components of a structure and construction of building elements in an off-site plant, which are subsequently moved and installed on the jobsite. In general, prefabrication methods can be primarily categorized into volumetric construction and panelization [68]. Volumetric construction involves the entire construction of building systems as individual box-like units off-site in a factory. These units are then conveyed and erected on-site. On the other hand, panelization entails the fabrication and assembly of the structural components of the building in the plant. The components are later moved to the jobsite to be installed [68].

1.2.3.4 Autonomous Equipment

In recent years, self-governing terrain vehicles and equipment have been the subject of extensive research and development across many industries. For instance, squadrons of large autonomous vehicles are already being implemented in the mining industry. However, the construction sector is just starting to explore this concept as many influential heavy equipment manufacturers have begun developing this technology [16, 34]. Komatsu, a large construction company, has developed a smart construction technology that employs UASs to survey and map out a construction jobsite in 3D and then uses the generated information to guide driverless equipment for excavation and earthmoving operations [34]. Other developments in autonomous construction equipment include Volvo construction equipment concept machines and Sandvik AutoMine products [30].

1.2.4 Intelligent Systems

1.2.4.1 Internet of Things (IoT)

IoT is a system of networks of devices that utilizes electronic gadgets like sensors and actuators to communicate information and bring data up to date to ensure the overall system works optimally [69]. In the construction industry, this structure is generally called telematics, a network of objects integrated with the systems of buildings and infrastructure to observe performance levels, working conditions, or physical states [70]. Information from the sensors can be incorporated into BIM during the operation phase of buildings. By utilizing the data, building occupancy trends such as energy use, people traffic, and movement or temperature preferences can be analyzed and modeled to enhance future projects [70]. Before this intelligent system can be widely adopted in the construction industry, it demands a robust connectivity infrastructure [69].

1.2.4.2 Artificial Intelligence (AI)

AI is the development of systems that capture and analyze large amounts of data to determine patterns and trends. AI utilizes the power of machines to model human intelligence, solve problems, and complete tasks with higher speed and accuracy [71]. AI has been employed for several construction applications, including building lifecycle management [72], predictions of construction project award price [73], optimized construction scheduling [74], accident prevention in construction zones [75], and prediction of formwork labor productivity [76].

1.2.4.3 Cloud Computing

Cloud computing is defined by researchers of the National Institute of Standard and Technology (NIST) as "a model for enabling ubiquitous, convenient, on-demand network access to a shared pool of configurable computing resources that can be rapidly provisioned and released with minimal management effort or service provider interaction" [77]. With cloud computing, unified services that store construction project information can be retrieved from the Internet for collaboration across the company and on the jobsite [16]. With cloud-based platforms, project team members can access project information from their devices connected to the Internet, enabling a file-sharing collaboration that allows them to view, manage, distribute, and collaborate in real time on project documents [16].

1.2.4.4 Big Data (BD)

Information is the essence of modern-day construction, and large volumes of data are generated daily on construction projects via BIM and other technologies. BD and predictive analysis process a vast volume of data and derive valuable acumen from the data pool [78]. With the need for project team members to continually manage the vast volume of data regularly collected on a project, BD solutions can ensure the right data is collected from all the data-generating devices implemented on a project [79].

1.2.4.5 Generative Design

The design of complex buildings and products requires designers and engineers to investigate numerous design and construction options. Generative design is an exploration process of establishing design constraints such as cost constraints, materials, performance or spatial requirements, and manufacturing methods and to come up with different design possibilities using computational analysis [80]. Through this process, all the possible permutations of a solution are explored, and multiple design alternatives are generated. Through generative design processes, lightweight structures can be designed and produced [81], and architectural and engineering designs can be consolidated by combining several parts into a single component [82]. Also, design solutions can be diversified by enabling designers to quickly explore design alternatives that would otherwise be overlooked [83]. Furthermore, together with AM, generative design can be used to design and manufacture high-performance products, including high-performance industrial robots and several robotic components that enhance the performance of the robot [84].

1.2.4.6 Communication and Tracking (CT) Systems

CT systems are wireless locating and tracking systems used for localizing, monitoring, and taking inventory of personnel or physical assets. A typical application is the use of radio-frequency identification (RFID) devices, radio-frequency (RF) tags, and radio-frequency data communication (RFDC) devices. RFID uses electromagnetic waves or vibrating magnetic fields to transfer signals to identification components connected to physical components [85]. With RFID, workers on the construction site can efficiently track and maintain job assets such as tools, machinery, materials, and prefabricated building elements to optimize the project cost and schedule [86]. Additionally, RFID can be employed for theft interception and inventory management of all job assets, as the embedded sensors can serve as a recovery mechanism for locating lost or stolen equipment [87]. Other applications of RFID on construction jobsites include personnel management and safety. A use case for the technology in these applications is to restrict the access of unauthorized workers on site and to keep track of the number and identifications of personnel who have access to the construction sites [87].

1.2.4.7 Wearable Sensors

Wearable sensors are technology incorporated with wearable pieces to monitor health conditions and provide relevant data for clinical care and health and safety monitoring and improvement [88]. The construction industry is considered one of the most dangerous and risky fields due to its high rate of work-related injuries and fatalities [89]. The high frequency of injuries and fatalities remains a major concern of construction industry practitioners and researchers. The adoption of wearable sensors such as gyroscopes, electrocardiogram (ECG/EKG) trackers, and physiological status monitoring (PSM) system integrated with GPS trackers can enable data collection and analysis to provide real-time health and safety results to construction personnel [90]. The collection and analysis of health and safety information are vital for measuring and improving safety strategy development, and wearable sensors can predict safety performance and inform management practices. Research has shown that wearable sensors used in other mature industries can monitor and measure a wide variety of safety performance metrics, including blood oxygen, blood pressure, heart rate, proximity detection, respiratory rate, and worker location tracking in order to prevent common injuries and fatalities on construction jobsites [90].

1.3 Automation and Robotics Technologies Deployment Trends in Construction

The previous section presented an account of the advancements in automation and robotics technologies from research [16, 32]. To understand the deployment of these technologies, semi-structured interviews were conducted with construction technology and innovation specialists to determine where each of the technologies stands in the current industrialization and digitization endeavors in the industry. Professionals from different organizations were selected and interviewed based on their position in the industry and comprised construction technology and innovation specialists, researchers, and educators.

The selection criteria focused on those who either directly worked with or managed automation and robotics technologies and championed research, integration, and training of these technologies in their organizations. A participant pool of 32 experts from industry and academia was created, and after initiating contact, 15 committed to participating in the study. The study participants consisted of five VDC professionals, two pre-construction technology managers, two construction technology and innovation managers, three researchers, and three associate professors. With semi-structured interviews, it was possible to collect substantial data, as the specialists were urged to analyze the subject matter and express their views based on their experiences with the technologies.

During the interviews, the specialists gave insights into the level of experience they had with these technologies by identifying the ones they had implemented in their construction workflows. Their opinions were analyzed based on the number of citations during the interviews. The specialists had varying degrees of experience with the different automation and robotics technologies being examined. Furthermore, these technologies are continuously evolving and becoming more advanced. To predict the future trends in automation and robotics technologies, the specialists were asked to identify the technologies they plan to implement in the next 5 and 10 years. The distributions of the different automation and robotics technologies that the specialists have incorporated in their construction workflows, and the projected technology trends are shown in Fig. 1.3.

The automation and robotics technologies with the top rates of current implementation were cloud computing (100%), laser scanners and UASs (80%), and 360-degree technology, VR, and AR (73%). The automation and robotics technologies with the least three rates of implementation were generative design (7%), wearable sensors and DT (13%), and AM, autonomous vehicles, and rovers (27%). However, the technologies with the least deployment rates are presently at the early adopter stage, and more than half of the specialists expect to be integrating the

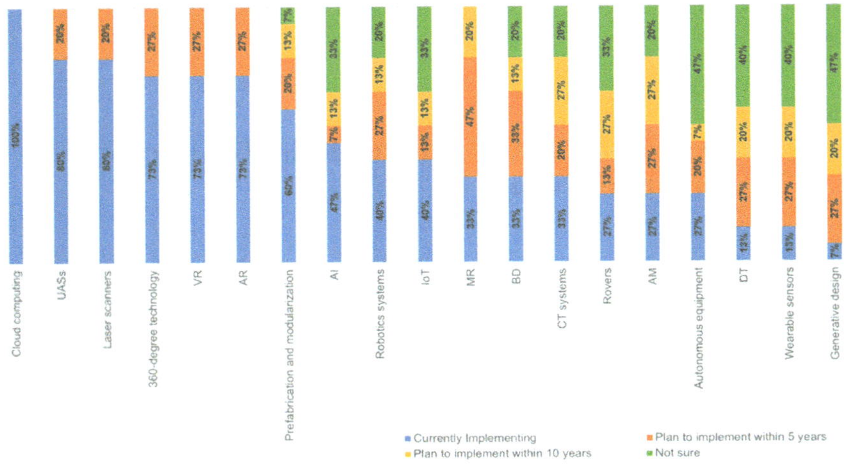

Fig. 1.3 Automation and robotics technologies deployment trends in construction

technologies in their construction workflows within 10 years. These identified projected trends give more insight into the automation and robotics technologies the specialists are most excited about using on future projects. They do not establish a definite trend indicating that the construction industry is moving toward automation and robotics.

1.4 Applications of Automation and Robotics Technologies in Construction

Significant efforts to automate the construction process and enhance efficiency in the industry have been prevalent since the nineteenth century, as seen in the various manifestations of large and technologically formidable construction like long-span bridges [91]. Construction activities that employed the use of machinery as an alternative to human labor were new modes of mechanization in construction. By the end of 1970, robotics equipment with the ability to lay masonry units were being utilized, and by the late 1980s, there was a growing demand for construction robots in Japan [7].

The range of application of construction automation and robotics technologies is extensive, spanning over all phases of the lifecycle of construction projects, from design through completion of construction [92]. These technologies can be implemented post-construction for maintenance and operation activities until eventual deconstruction. However, the level of automation and robotics technologies implementation differs significantly throughout the different stages of construction. For instance, the automation of design in construction by BIM implementation is prevalent nowadays; on the other hand, the employment of construction automation and robotics equipment for operations on the jobsite is not commonplace [68].

1.4.1 Construction Process Automation

Automation and robotics technologies have been deployed to partially or fully automate construction processes and activities, which are usually manual and competitive. By automating construction processes, companies can increase productivity, efficiency, and quality on projects. When implemented correctly, process automation could eliminate the current labor shortage facing the industry [93]. Automation and robotics technologies have been developed to perform autonomous construction tasks, both on-site and off-site, as they can be programmed with customized abilities to manipulate materials and equipment [6]. Robotic systems have been developed to tackle tasks, including glazing installation, drywalling, painting, jointing, concrete finishing, welding, and drilling. The capabilities of the technology concerning reach, load capacity, and the number of degrees of freedom result in improved efficiency of work processes [93].

3D printing robots have also been deployed for building large-scale masonry, concrete, and steel structures [35]. However, a common limitation of these robots is their large sizes. UASs have recently become attractive for additive construction due to their swarm behavior attributed to their small sizes. Unlike large-scale robotic systems, UASs can navigate construction sites without human intervention [94]. Researchers have explored the use of UASs in building real-scale masonry structures, as they facilitate construction in places that are difficult to reach without the use of cranes, which reduces congestion on the jobsite [95]. The study results demonstrated reduced schedule, decreased work defects, lesser impact of human factors, lessening of harmful job tasks, increased safety, and increased profitability.

1.4.2 On-Site Fabrication and Operations

Automation and robotics technologies can be deployed directly on jobsites to complete construction tasks. Robotic systems have been developed to complete single tasks repetitively, such as brick assembly, steel erection, welding, wall façade installation, painting, and concrete spraying [96]. The use of these systems for on-site operations can economically condense construction schedules and improve on-site safety as they complex tasks that are otherwise repetitive and unsafe [30]. The STCR systems approach could be customized to be operated together with other conventional construction techniques and equipment. However, because of the dynamic nature of construction sites and the complexity of integrating with other human-centered activities, these robotic systems could prove challenging to deploy on-site [96].

Automated and robotic on-site factories, which integrate several STCRs into structured environments on jobsites, have been developed to address the limitation of stand-alone robots [30]. Researchers have performed studies to support cooperation between multiple robots to enable the completion of more difficult operations. For instance, [97] developed a 3D printing system that employs multiple mobile robots to print a single, large, concrete structure collaboratively. The greater flexibility of the robotic arms allowed for the 3D printers to produce complex curved elements. The swarm printing allowed for higher productivity and scalability, more efficient mechanical stability, and alleviated transportation costs [97].

1.4.3 Off-Site Construction

Off-site construction involves the automated fabrication and assembly of construction elements in a controlled factory at a location to support the rapid and efficient construction of a structure at the final installation location [98]. Many automation and robotics technologies support off-site fabrication and assembly, including prefabrication and modularization, 3D printing, robotic systems, RFID, IoT, AI BD, laser scanners, VR, and AR [99]. Off-site construction is often synonymous with prefabricated construction and modular construction, both of which involve building component manufacturing (BCM) approaches and also large-scale prefabrication (LSP) approaches [10].

3D printing techniques have also been deployed for off-site construction operations, especially for generating complex, custom-built geometries for building elements and reduced reliance on formwork and other associated temporary structures used during construction [100]. Off-site construction alleviates several problems associated with traditional methods. By moving a large proportion of construction work away from an uncontrolled, open-air environment to a controlled indoor factory setting with round-the-clock operational time potential, off-site construction offers enhanced worker safety, decreased labor costs, and reduced material waste. Off-site construction provides a solution to the shortage of skilled labor and replaces traditional supply chains [100].

1.4.4 Earthwork Operations

As the construction industry advances into the era of industrialization, there has been an increase in the deployment of autonomous vehicles and heavy equipment on jobsites. Autonomous equipment has been used for earthwork and earthmoving operations to decrease operation costs, enhance productivity, improve automation, and increase safety [101]. Heavy equipment such as bulldozers, wheel loaders, excavators, and motor graders have several blind spots that increase the possibility of struck-by collisions with workers or dangerous utilities. Pedestrian-vehicular incidents, where a worker was struck by a vehicle or mobile heavy equipment, were the most common cause of road construction fatalities between 2011 and 2016, resulting in 267 deaths [102]. In 2019, there were 256 reported transportation incidents in construction, including fatal occupational injuries resulting from being struck by a vehicle [103].

One of the most important advantages of autonomous equipment is that they provide a safer jobsite. Some of the emerging technologies incorporated into autonomous equipment include tracking, guidance, and proximity detection systems to prevent accidents [101]. Built robotics is at the forefront of developing autonomous equipment for construction operations, with an autonomous equipment portfolio that includes autonomous dozer, excavator, and compact track loader [104]. Other researchers have evaluated the automation and remote control of earthmoving equipment such as wheel loaders; load-haul-dump (LHD) machines; and excavators to improve the efficiency, quality, and safety of existing earthwork processes [105, 106].

1.4.5 Jobsite Surveying and Construction Progress Monitoring

Construction jobsite surveying and progress monitoring are traditionally very monotonous tasks that require capturing, organizing, and interpreting a vast volume of site photos and videos regularly. Automation and robotics technologies, such as 360-degree cameras, UASs, rovers, laser scanners, and robotic total stations, can help speed up construction surveying processes and improve construction data quality and accuracy. UASs, outfitted with sensors such as multispectral cameras, 360-degree cameras, thermal imaging cameras, and LiDAR scanners, have been used to survey jobsites and surrounding areas before construction commences and monitor project progress throughout construction [35, 107].

The construction data captured with the UAS can be compared against BIM data to assess the project progress and check for compliance with geometric design models through enhanced clash detection [108]. The captured data can also be used for energy analysis, health and safety evaluation, and potential hazard identification [35]. As traditional site surveys can be very time-consuming, deploying automation and robotics technologies speeds up the completion of the task and reduces dependence on human resources.

1.4.6 Inspection and Maintenance

Regular inspection and maintenance of buildings and structures are crucial for ensuring operational efficiency at all times. Buildings and structures are sometimes inspected less frequently than optimal due to inspection costs, length of procedures, interruptions of routine operations, and risk of complex and unsafe conditions [109].

Automation and robotics technologies such as micro aerial vehicles, robotic systems, UASs, and autonomous vehicles can be equipped with cameras and sensors and can be employed to complete routine inspection and maintenance of buildings and infrastructure [109].

The application of automation and robotics technologies in inspection and maintenance include automated bridge inspection and reporting [110], detection and localization of surface distress in unpaved roads [111], compliance inspection of runway design codes for airfields [112], and decision-making support during the inspection process [113]. Deploying these technologies for inspection and maintenance operations minimizes risk in dangerous inspection environments, enhances the efficiency of inspection processes, lessens subjectivity, and improves the accuracy of the inspection process [109].

1.4.7 Health and Safety Monitoring

Construction jobsites are continuously changing environments, often involving risky and dangerous working conditions resulting in high-site-related injuries and fatalities. The collection and assessment of safety information are crucial for measuring and improving health and safety policies [90]. There is a need for construction jobsites to be monitored continuously to identify hazardous work conditions and protect workers from possible injuries and deadly events. Typically, construction health and safety monitoring procedures rely greatly on manual inspection, which is prone to human error [114].

Automation and robotics technologies can be implemented to reduce labor-intensive efforts associated with construction health and safety monitoring. UASs, wearable sensors, and other sensor-based technologies are implemented to ensure construction jobsite safety [115]. UASs have been developed to automatically collect, identify, and assess fall hazards in construction projects [116]. Applications of wearable sensors in construction include situational awareness, environmental conditions detection, and detection of workers' proximity to danger zones [117, 118].

1.4.8 Real-Time Decision-Making

Construction processes are typically dynamic and unstructured and require multi-discipline collaboration, which creates uncertainties and logistical challenges for automated decision-making in recurring construction tasks [119]. With the increasing emergence of massive uncertainty datasets, data-driven optimization paves a new way to decision-making under uncertainty [120]. Automation and robotics technologies can be leveraged to inform real-time decision-making on the construction site. By deploying IoT infrastructure, decision process models can be developed to optimize construction tasks and processes. One application is to model the dynamics of the construction environment in response to several variables for use in decision analysis and planning, where the goal is to identify the best sequence of actions to take to optimize the project outcomes [119].

Furthermore, a data-driven optimization framework powered by machine learning and BD aims to improve the understanding of construction data patterns. The outcomes are utilized in solving crucial and challenging problems such as the prediction of outcomes or safety hazard detection [120]. Data-driven process optimization tackles the unpredictable and intricate nature of construction processes. Also, the transference of a computer-generated model into the real environment makes it possible to interpret the model results without human interaction leading to improved scientific planning and control approaches in project management [119]. Overall, a practical implementation of IoT infrastructure supplies opportune and actionable insights through real-time and automated monitoring and control.

1.4.9 Inventory and Stock Management

Inventory and stock management in construction involve the procurement, identification, transport, and storage of materials on projects [121]. The proper management of supply levels of construction materials and equipment availability is fundamental to guarantee that the project is completed on time and within budget. The lack of proper handling and storage of materials on jobsites has made it difficult to track and locate material availability [122]. There has been an increase in the integration of automation and robotics technologies, such as RFID devices, to support the identification and location of materials and equipment to enhance the just-in-time delivery of construction materials [93]. The benefits of an integrated tracking system include reduced project delays, reduced costs, and increased efficiency.

1.4.10 Demolition, Material Sorting, and Recycling

Automation and robotics technologies have been implemented in the industry to improve the environmental performance of construction processes in the context of sustainability. In 2018, 600 million tons of construction and demolition waste were generated, with demolition representing more than 90% of the total waste generated [123]. Automation and robotics technologies can be used during demolition to optimize the use and reuse of materials to minimize waste [124]. There are different types of demolition robotics technologies currently available in the market or being developed. Demolition robots can perform deconstruction tasks much more precisely and efficiently than manual labor and handheld equipment [21]. These robots can be used to disassemble building components systematically. For instance, heavy steel elements can be carefully taken apart without being melted or brought down using explosives, thereby preserving their condition during recycling [124].

Automated construction processes can foster the efficient use of materials on projects, resulting in reduced waste during construction and demolition. With CT systems, sensor-based control can be used to track materials and other building components during construction and eventually during demolition [93]. By combining BIM and automation and robotics technologies, sensors can detect the geometry of waste materials and components for reuse [125]. Automation and robotics technologies can result in savings through the recycling of construction and demolition waste. They also improve the health and safety of construction workers and jobsites as they prevent air pollution and protect workers from proximity to hazardous waste [21]. The major negative impact of demolition robots is that they require the additional consumption of a significant amount of energy for operation [124].

1.5 Conclusions

The construction industry is swiftly transforming in the era of industrialization, and innovative technologies as automation and robotics technologies are altering the nature of jobsite tasks and processes. In construction, automation and robotics is defined as the application of autonomous machines that carry out construction tasks and operations automatically through intelligent programming and controls. These technologies are implemented in the industry to automate construction processes and enhance efficiency and productivity. Nineteen automation and robotics technologies were identified and categorized into four technology groups: reality computing, visualization, automation, and intelligent systems.

This study explored the extensive range of applications of automation and robotics technologies in construction which typically spans across the lifecycle of projects and includes documentation of existing site conditions and inspections, design feasibility and coordination, site progress monitoring, and health and safety evaluation and training, proposals, and project pursuits and visualization. Most of the identified applications require human intervention and human-robot collaboration, with a few applications involving total automation.

Since construction is one of the least digitized industries, automation and robotics technologies will have a significant impact on its productivity. Therefore, as construction companies increasingly and successfully adopt these technologies to automate tasks and processes, the market for these automation and robotics technology will grow steadily. This chapter lays the foundation for promoting the integration of automation and robotics technologies in the industry by presenting a comprehensive review and analysis of existing research, with the aim of identifying the opportunities for future applications in construction processes.

References

1. F. Barbosa, J. Woetzel, J. Mischke, M.J. Ribeirinho, M. Sridhar, N. Bertram, S. Brown, *Reinventing Construction: A Route to Higher Productivity* (McKinsey Global Institute, McKinsey & Company, 2017) https://www.mckinsey.com/business-functions/operations/our-insights/reinventing-construction-through-a-productivity-revolution. Accessed 10.03.17
2. S. Witthoeft, I. Kosta, *Shaping the Future of Construction. Inspiring Innovators Redefine the Industry* (World Economic Forum, 2017) http://www3.weforum.org/docs/WEF_Shaping_the_Future_of_Construction_Inspiring_Innovators_redefine_the_industry_2017.pdf. REF 020117. Accessed 12.04.18
3. United Nations, World Urbanization Prospects 2018, illustrated revised ed., United Nations Publications, UN, 2019. ISBN 9211483190, 9789211483192
4. J. Woetzel, N. Garemo, J. Mischke, P. Kamra, R. Palter, *Bridging Infrastructure Gaps. Has the World Made Progress?* (McKinsey Global Institute, McKinsey & Company, 2017) https://www.mckinsey.com/business-functions/operations/our-insights/bridging-infrastructure-gaps-has-the-world-made-progress. Accessed 13.01.18
5. Bureau of Labor Statistics, Integrated BLS-BEA Industry-Level Production Account Tables. https://www.bls.gov/mfp/indprod11.xlsx, 2018. Accessed 02.27.2021
6. T. Bock, T. Linner, *Robot-Oriented Design: Design and Management Tools for the Deployment of Automation and Robotics in Construction* (Cambridge University Press, Cambridge, 2015). https://doi.org/10.1017/cbo9781139924146
7. S.S. Kamaruddin, M.F. Mohammad, R. Mahbub, Barriers and impact of mechanisation and automation in construction to achieve better quality products. Procedia. Soc. Behav. Sci. **222**, 111–120 (2016). https://doi.org/10.1016/j.sbspro.2016.05.197
8. T. Linner, W. Pan, R. Hu, C. Zhao, K. Iturralde, M. Taghavi, J. Trummer, M. Schlandt, T. Bock, A technology management system for the development of single-task construction robots. Constr. Innov. **20**, 96–111 (2020). https://doi.org/10.1108/CI-06-2019-0053
9. R. Agarwal, S. Chandrasekaran, M. Sridhar, *Imagining construction's digital future* (McKinsey Productivity Sciences Center, Singapore, McKinsey & Company, 2016) https://www.mckinsey.com/business-functions/operations/our-insights/imagining-constructions-digital-future. Accessed 15.08.19
10. T. Bock, The future of construction automation: Technological disruption and the upcoming ubiquity of robotics. Automat. Constr. **59**, 113–121 (2015). https://doi.org/10.1016/j.autcon.2015.07.022
11. Autodesk, Inc. 2018, Connected BIM for building design: Improved project insight with the cloud. https://www.autodesk.com/solutions/bim/discover-building-design/bim-for-building-design. Accessed 15.06.19
12. S. Duvernoy, Architecture's new media: Principles, theories, and methods of computer-aided design. Nexus Netw. J. **7**, 163–165 (2005). https://doi.org/10.1007/s00004-005-0015-1
13. R. Sacks, C. Eastman, G. Lee, P. Teicholz, *BIM Handbook A Guide to Building Information Modeling for Owners, Designers, Engineers, Contractors, and Facility Managers*, 3rd edn. (Wiley, 2018) ISBN: 978-1-119-28753-7
14. A. Krivonogov, G. Zakharova, S. Kruglikov, S. Plotnikov, Implementation of BIM-technologies in the educational program of the architectural university. MATEC Web Conf. **146**, 01001 (2018). https://doi.org/10.1051/matecconf/201814601001
15. K. Schwab, The fourth industrial revolution (industry 4.0) a social innovation perspective. Techn. Innov. Manag. Rev. **7**, 12–20 (2017). https://doi.org/10.22215/timreview/1117
16. T.D. Oesterreich, F. Teuteberg, Understanding the implications of digitisation and automation in the context of Industry 4.0: A triangulation approach and elements of a research agenda for the construction industry. Comput. Ind. **83**, 121–139 (2016). https://doi.org/10.1016/j.compind.2016.09.006
17. Merriam-Webster, *Robot | Definition of Robot by Merriam-Webster* (Merriam-Webster, 2018) https://www.merriam-webster.com/dictionary/robot. Accessed 10.10.18

18. R. Rennie, J. Law (eds.), *A Dictionary of Physics*, 8th edn. (Oxford University Press, 2019). https://doi.org/10.1093/acref/9780198821472.001.0001
19. M. Hewitt, J. Gambatese, Automation Consideration During Project Design, in *Proceedings of the 19th International Symposium on Automation and Robotics in Construction*, (The International Association for Automation and Robotics in Construction, Washington, USA, 2002), pp. 197–203. https://doi.org/10.22260/isarc2002/0031
20. S.M.S. Elattar, Automation and robotics in construction: Opportunities and challenges, Emirates. J. Eng. Res. **13**, 21–26 (2008) http://citeseerx.ist.psu.edu/viewdoc/summary?doi=10.1.1.524.7909
21. A.N. Ruggiero, C.L.S. Laurent, S.D. Salvo, *Robotics in Construction* (Worcester Polytechnic Institute Digital WPI, 2016) https://digitalcommons.wpi.edu/iqp-all/2749?utm_source=digitalcommons.wpi.edu%2Fiqp-all%2F2749&utm_medium=PDF&utm_campaign=PDFCoverPages. Accessed 20.03.19
22. A. Kerr (ed.), *Dictionary of World History*, 3rd edn. (Oxford University Press, Oxford, 2015). https://doi.org/10.1093/acref/9780199685691.001.0001
23. World Encyclopedia (Phillip's, Oxford, 2014). https://doi.org/10.1093/acref/9780199546091.001.0001
24. Merriam-Webster, *Automation | Definition of Automation by Merriam-Webster* (Merriam-Webster, 2018) https://www.merriam-webster.com/dictionary/automation. Accessed 10.10.18
25. J. Scott, G. Marshall, *A Dictionary of Sociology*, 3rd edn. (Oxford University Press, Oxford, 2009). https://doi.org/10.1093/acref/9780199533008.001.0001
26. M. Rajgor, J. Pitroda, Automation: A new millennium technology for construction industries. Global Res. Anal. **9**, 79–81 (2013) ISSN 2277-8160
27. C. Soanes, A. Stevenson (eds.), *Concise Oxford English Dictionary*, 11th edn. (Oxford University Press, Oxford, 2005) ISBN-10: 019861-475
28. Merriam-Webster, *Construction | Definition of Construction by Merriam-Webster* (Merriam-Webster, 2018) https://www.merriam-webster.com/dictionary/construction. Accessed 10.10.18
29. R.A. Buswell, R.C. Soar, A.G.F. Gibb, A. Thorpe, Freeform construction: Mega-scale rapid manufacturing for construction. Automat. Constr. **16**, 224–231 (2007). https://doi.org/10.1016/j.autcon.2006.05.002
30. T. Bock, T. Linner, *Construction Robots: Elementary Technologies and Single-Task Construction Robots* (Cambridge University Press, Cambridge, 2017). https://doi.org/10.1017/CBO9781139872041
31. B. Jørgensen, S. Emmitt, Lost in transition: The transfer of lean manufacturing to construction. Eng. Const. Archit. Manage. **15**, 383–398 (2008). https://doi.org/10.1108/09699980810886874
32. L.D. Xu, E.L. Xu, L. Li, Industry 4.0: State of the art and future trends. Int. J. of Prod. Res. **56**, 2941–2962 (2018). https://doi.org/10.1080/00207543.2018.1444806
33. J.M. Beer, A.D. Fisk, W.A. Rogers, Toward a framework for levels of robot autonomy in human-robot interaction. J Hum Robot Interact. **3**, 74–99 (2014). https://doi.org/10.5898/JHRI.3.2.Beer
34. H. Shakhatreh, A.H. Sawalmeh, A. Al-Fuqaha, Z. Dou, E. Almaita, I. Khalil, N.S. Othman, A. Khreishah, M. Guizani, Unmanned aerial vehicles (UAVs): A survey on civil applications and key research challenges. IEEE Access. **7**, 48572–48634 (2019). https://doi.org/10.1109/ACCESS.2019.2909530
35. R. Bogue, What are the prospects for robots in the construction industry? Ind. Rob. **45**, 1–6 (2018). https://doi.org/10.1108/IR-11-2017-0194
36. J. Chen, Y. Fang, Y.K. Cho, C. Kim, Principal axes descriptor for automated construction-equipment classification from point clouds. J. Comput. Civ. Eng. **31**, 04016058 (2017). https://doi.org/10.1061/(asce)cp.1943-5487.0000628
37. E. Fallon (ed.), *A Look at 3D Laser Scanning in the Construction Industry and Beyond, The Expert Report: The Newsletter of the Duggan Rhodes Group* (Pittsburg, PA, 2012)

38. T. Randall, Construction engineering requirements for integrating laser scanning technology and building information modeling. J. Constr. Eng. Manag. **137**, 797–805 (2011). https://doi.org/10.1061/(asce)co.1943-7862.0000322
39. S. Chiodini, R.G. Reid, B. Hockman, I.A.D. Nesnas, S. Debei, M. Pavone, Robust visual localization for hopping rovers on small bodies, in *2018 IEEE International Conference on Robotics and Automation*, (Institute of Electrical and Electronics Engineers, Brisbane, QLD, Australia, 2019), pp. 897–903. https://doi.org/10.1109/ICRA.2018.8462865
40. T. Narumi, S. Aoki, F. Muramatsu, Indoor Visualization Experiments at Building Construction Site Using High Safety UAV, in *Proceedings of the 36th International Symposium on Automation and Robotics in Construction*, (The International Association for Automation and Robotics in Construction, Banff, Canada, 2019), pp. 961–966. https://doi.org/10.22260/isarc2019/0128
41. Doxel, Artificial intelligence for construction productivity (2020). https://www.doxel.ai. Accessed 10.09.19
42. R. Eiris, M. Gheisari, B. Esmaeili, PARS: Using augmented 360-degree panoramas of reality for construction safety training. Int. J. Environ. Res. Public Health **15**, 2452 (2018). https://doi.org/10.3390/ijerph15112452
43. N.J. Shih, J.T. Lai, Y.L. Tsai, The application of a panorama image database management systems (PIDMS) for information integration on construction sites. ITcon. **11**, 641–654 (2006) https://www.itcon.org/2006/44
44. M. Gheisari, G. Williams, Using Augmented Panoramic Views as an Online Course Delivery Mechanism in MOOCs, in *51st ASC Annual International Conference Proceedings*, (Associated Schools of Construction, College Station, TX, 2015) http://ascpro0.ascweb.org/archives/cd/2015/paper/CERT337002015.pdf
45. R. Eiris, H.I. Moud, M. Gheisari, Using 360-Degree interactive panoramas to develop virtual representation of construction sites, in *Lean and Computing in Construction Congress*, (ITC Digital Library, Heraklion, Greece, 2017), pp. 775–782. https://doi.org/10.24928/jc3-2017/0122
46. H.C. Pham, N.N. Dao, A. Pedro, Q.T. Le, R. Hussain, S. Cho, C.S. Park, Virtual field trip for mobile construction safety education using 360-degree panoramic virtual reality. Int. J. Eng. Educ. **34**, 1174–1191 (2018) https://www.ijee.ie/latestissues/Vol34-4/05_ijee3626.pdf
47. K. Kensek, D. Noble, M. Schiler, A. Tripathi, Augmented reality: An application for architecture, in *Eighth International Conference on Computing in Civil and Building Engineering*, (American Society of Civil Engineers, Stanford, CA, 2000), pp. 294–301. https://doi.org/10.1061/40513(279)38
48. J. Fogarty, J. McCormick, S. El-Tawil, Improving student understanding of complex spatial arrangements with virtual reality. J. Prof. Issues in Eng. Educ. Pract. **144**, 04017013 (2018). https://doi.org/10.1061/(ASCE)EI.1943-5541.0000349
49. F.M. Bademosi, R. Tayeh, R.R.A. Issa, An Immersive Approach to Construction Cost Estimating, in *ASCE International Conference on Computing in Civil Engineering 2019*, (American Society of Civil Engineers, Atlanta, GA, 2019), pp. 48–54. https://doi.org/10.1061/9780784482421.007
50. M. Froehlich, S. Azhar, *Investigating Virtual Reality Headset Applications in Construction, in: 52nd ASC Annual International Conference Proceedings* (Associated Schools of Construction, Provo, UT, 2016) http://ascpro0.ascweb.org/archives/cd/2016/paper/CPRT195002016.pdf
51. R.K. Soman, J.K. Whyte, A framework for cloud-based virtual and augmented reality using real-time information for construction progress monitoring, in *Lean and Computing in Construction Congress*, (ITC Digital Library, Heraklion, Greece, 2017), pp. 833–840. https://doi.org/10.24928/jc3-2017/0273
52. A. Osello, G. Lucibello, F. Morgagni, HBIM and virtual tools: A new chance to preserve architectural heritage. Buildings **8**, 12 (2018). https://doi.org/10.3390/buildings8010012
53. Y. Mo, D. Zhao, J. Du, W. Liu, A. Dhara, Data-driven approach to scenario determination for VR-based construction safety training, in *Construction Research Congress 2018*,

(Americana Society of Civil Engineers, New Orleans, LA, 2018), pp. 114–125. https://doi.org/10.1061/9780784481288.012
54. T.H. Lin, C.H. Liu, M.H. Tsai, S.C. Kang, Using augmented reality in a multiscreen environment for construction discussion. J. Comput. Civ. Eng. **29**, 04014088 (2015). https://doi.org/10.1061/(asce)cp.1943-5487.0000420
55. P. Milgram, H. Takemura, A. Utsumi, F. Kishino, Augmented reality: A class of displays on the reality-virtuality continuum, in *Photonics for Industrial Applications*, (SPIE Digital Library, Boston, MA, 1995), pp. 282–292. https://doi.org/10.1117/12.197321
56. A.B. Cleveland, Emerging tools to enable construction engineering. J. Constr. Eng. Manag. **137**, 836–842 (2011). https://doi.org/10.1061/(asce)co.1943-7862.0000278
57. C.S. Park, D.Y. Lee, O.S. Kwon, X. Wang, A framework for proactive construction defect management using BIM, augmented reality and ontology-based data collection template. Automat. Constr. **33**, 61–71 (2013). https://doi.org/10.1016/j.autcon.2012.09.010
58. S. Rankohi, L. Waugh, Review and analysis of augmented reality literature for construction industry. Vis. Eng. **1**, 9 (2013). https://doi.org/10.1186/2213-7459-1-9
59. X. Wang, P.E.D. Love, M.J. Kim, C.S. Park, C.P. Sing, L. Hou, A conceptual framework for integrating building information modeling with augmented reality. Automat. Constr. **34**, 37–44 (2013). https://doi.org/10.1016/j.autcon.2012.10.012
60. F.M. Bademosi, R.R.A. Issa, Business Value of Augmented Reality in the Construction Industry, in: Construction Research Congress 2018, American Society of Civil Engineers, New Orleans, LA, 2018: pp. 715–724. https://doi.org/10.1061/9780784481264.070
61. F.M. Bademosi, N. Blinn, R.R.A. Issa, Use of augmented reality technology to enhance comprehension of construction assemblies. ITcon. **24**, 58–79 (2019) http://www.itcon.org/2019/4
62. P.S. Dunston, X. Wang, Mixed reality-based visualization interfaces for architecture, engineering, and construction industry. J. Constr. Eng. Manag. **131**, 1301–1309 (2005). https://doi.org/10.1061/(asce)0733-9364(2005)131:12(1301)
63. H.K. Wu, S.W.Y. Lee, H.Y. Chang, J.C. Liang, Current status, opportunities and challenges of augmented reality in education. Comput. Educ. **62**, 41–49 (2013). https://doi.org/10.1016/j.compedu.2012.10.024
64. V. Stojanovic, M. Trapp, R. Richter, B. Hagedorn, J. Döllner, Towards the Generation of Digital Twins for Facility Management Based on 3D Point Clouds, in *Proceedings of the 34th Annual ARCOM Conference*, (Association of Researchers in Construction Management, Belfast, UK, 2018), pp. 270–279. http://www.arcom.ac.uk/-docs/proceedings/b65e593d342a8de045cf05698677e600.pdf
65. T. Greif, N. Stein, C.M. Flath, Peeking into the void: Digital twins for construction site logistics. Comput. Ind. **121**, 103264 (2020). https://doi.org/10.1016/j.compind.2020.103264
66. C. Kan, C.J. Anumba, Digital Twins as the Next Phase of Cyber-Physical Systems in Construction, in *International Conference Computing in Civil Engineering 2019*, (American Society of Civil Engineers, Atlanta, GA, 2019), pp. 256–264. https://doi.org/10.1061/9780784482438.033
67. F. Bos, R. Wolfs, Z. Ahmed, T. Salet, Additive manufacturing of concrete in construction: Potentials and challenges of 3D concrete printing. Virtual Phys. Prototy. **11**, 209–225 (2016). https://doi.org/10.1080/17452759.2016.1209867
68. J. Neelamkavil, Automation in the Prefab and Modular Construction Industry, in *Proceedings of the 26th International Symposium on Automation and Robotics in Construction*, (The International Association for Automation and Robotics in Construction, Austin, TX, 2009), pp. 299–306. https://doi.org/10.22260/isarc2009/0018
69. H.N. Saha, A. Mandal, A. Sinha, Recent Trends in the Internet of Things, in *IEEE 7th Annual Computing and Communication Workshop and Conference*, (Institute of Electrical and Electronics Engineers, Las Vegas, NV, 2017), pp. 1–4. https://doi.org/10.1109/CCWC.2017.7868439
70. S. Greengard, *The Internet of Things* (MIT Press, Boston, 2015) ISBN: 9780262527736

71. M. Negnevitisky, *Artificial Intelligence A Guide to Intelligent Systems*, 2nd edn. (Pearson Education Limited, Essex, 2005) ISBN: 0 321 20466 2
72. L. Ustinovičius, R. Rasiulis, L. Nazarko, T. Vilutiene, M. Reizgevicius, Innovative research projects in the field of building lifecycle management. Procedia Eng. **122**, 166–171 (2015). https://doi.org/10.1016/j.proeng.2015.10.021
73. J.S. Chou, C.W. Lin, A.D. Pham, J.Y. Shao, Optimized artificial intelligence models for predicting project award price. Automat. Constr. **54**, 106–115 (2015). https://doi.org/10.1016/j.autcon.2015.02.006
74. V. Faghihi, A. Nejat, K.F. Reinschmidt, J.H. Kang, Automation in construction scheduling: A review of the literature. Int. J. Adv. Manuf. Technol. **81**, 1845–1856 (2015). https://doi.org/10.1007/s00170-015-7339-0
75. D. Nozaki, K. Okamoto, T. Mochida, X. Qi, Z. Wen, S.H. Myint, K. Tokuda, T. Sato, K. Tamesue, AI Management System to Prevent Accidents in Construction Zones Using 4K Cameras Based on 5G Network, in *Proceedings of the 21st International Symposium on Wireless Personal Multimedia*, (Institute of Electrical and Electronics Engineers, Chiang Rai, Thailand, 2018), pp. 462–466. https://ieeexplore.ieee.org/document/8712896/
76. S. Golnaraghi, Z. Zangenehmadar, O. Moselhi, S. Alkass, A.R. Vosoughi, Application of artificial neural network(s) in predicting formwork labour productivity. Adv. Civ. Eng. **2019**, 5972620 (2019). https://doi.org/10.1155/2019/5972620
77. P. Mell, T. Grance, Appendix C - The NIST definition of cloud computing, in: R. Strum, C. Pollard, J. Craig (eds), Application Performance Management (APM) in the Digital Enterprise, Kauffman, 2017: pp. 267–269. https://doi.org/10.1016/b978-0-12-804018-8.15003-x
78. M. Bilal, L.O. Oyedele, O.O. Akinade, S.O. Ajayi, H.A. Alaka, H.A. Owolabi, J. Qadir, M. Pasha, S.A. Bello, Big data architecture for construction waste analytics (CWA): A conceptual framework. J. Build. Eng. **6**, 144–156 (2016). https://doi.org/10.1016/j.jobe.2016.03.002
79. J. McMalcolm, *How Big Data Is Transforming the Construction Industry* (Construction Global, 2015) https://www.constructionglobal.com/technology-and-ai-1/how-big-data-is-transforming-the-construction-industry. Accessed 10.10.19
80. S. Krish, A practical generative design method. Comput. Aided Des. **43**, 88–100 (2011). https://doi.org/10.1016/j.cad.2010.09.009
81. E. Rodrigues, N. Soares, M.S. Fernandes, A.R. Gaspar, Á. Gomes, J.J. Costa, An integrated energy performance-driven generative design methodology to foster modular lightweight steel framed dwellings in hot climates. Energy Sustain. Dev. **44**, 21–36 (2018). https://doi.org/10.1016/j.esd.2018.02.006
82. P. Tang, X. Wang, X. Shi, Generative design method of the facade of traditional architecture and settlement based on knowledge discovery and digital generation: A case study of Gunanjie Street in China. Int. J of Archit. Herit. **13**, 679–690 (2019). https://doi.org/10.1080/15583058.2018.1463415
83. E. Touloupaki, T. Theodosiou, Energy performance optimization as a generative design tool for nearly zero energy buildings. Proc. Eng. **180**, 1178–1185 (2017). https://doi.org/10.1016/j.proeng.2017.04.278
84. E. Francalanza, A. Fenech, P. Cutajar, Generative design in the development of a robotic manipulator. Proc. CIRP **67**, 244–249 (2018). https://doi.org/10.1016/j.procir.2017.12.207
85. K.B. Sørensen, P. Christiansson, K. Svidt, Ontologies to support RFID-based link between virtual models and construction components. Comput. Aided Civ. Infrastr. Eng. **25**, 285–302 (2010). https://doi.org/10.1111/j.1467-8667.2009.00638.x
86. J. Majrouhi Sardroud, Influence of RFID technology on automated management of construction materials and components. Scien. Iranica. **19**, 381–392 (2012). https://doi.org/10.1016/j.scient.2012.02.023
87. M. Roberti, *How is RFID being used in the construction industry?* (Ask Experts Forum, RFID Journal, 2013) https://www.rfidjournal.com/question/how-is-rfid-being-used-in-the-construction-industry. Accessed 08.09.19

88. F. Fang, S. Aabith, S. Homer-Vanniasinkam, M.K. Tiwari, High-resolution 3D printing for healthcare underpinned by small-scale fluidics. 3D Print Med., 167–206 (2017). https://doi.org/10.1016/B978-0-08-100717-4.00023-5
89. OSHA – Occupational Safety and Health Administration, *Commonly Used Statistics* (OSHA Training Institute, 2018) https://www.osha.gov/data/commonstats. Accessed 01.02.19
90. I. Awolusi, E. Marks, M. Hallowell, Wearable technology for personalized construction safety monitoring and trending: Review of applicable devices. Automat. Constr. **85**, 96–106 (2018). https://doi.org/10.1016/j.autcon.2017.10.010
91. J.A. Romero, A.A. Lozano-Guzmán, E. Betanzo-Quezada, C.S. López-Cajún, Robotics and road transportation: A review, in *Intelligent Robotics and Applications, ICIRA 2014, Lecture Notes in Computer Science*, ed. by X. Zhang, H. Liu, Z. Chen, N. Wang, vol. 8917, (Springer, Cham, 2014). https://doi.org/10.1007/978-3-319-13966-1_46
92. K.S. Saidi, T. Bock, C. Georgoulas, Robotics in construction, in *Springer Handbook of Robotics*, Springer Handbooks, ed. by B. Siciliano, O. Khatib, (Springer, Cham, 2016). https://doi.org/10.1007/978-3-319-32552-1_57
93. Q. Chen, B. García de Soto, B.T. Adey, Construction automation: Research areas, industry concerns and suggestions for advancement. Automat. Constr. **94**, 22–38 (2018). https://doi.org/10.1016/j.autcon.2018.05.028
94. J. Werfel, K. Petersen, R. Nagpal, Designing collective behavior in a termite-inspired robot construction team. Science **342**, 754–758 (2014). https://doi.org/10.1126/science.1245842
95. S. Goessens, C. Mueller, P. Latteur, Feasibility study for drone-based masonry construction of real-scale structures. Automat. Constr. **94**, 458–480 (2018). https://doi.org/10.1016/j.autcon.2018.06.015
96. J.M. Davila Delgado, L. Oyedele, A. Ajayi, L. Akanbi, O. Akinade, M. Bilal, H. Owolabi, Robotics and automated systems in construction: Understanding industry-specific challenges for adoption. J. Build. Eng. **26**, 100868 (2019). https://doi.org/10.1016/j.jobe.2019.100868
97. X. Zhang, M. Li, J.H. Lim, Y. Weng, Y.W.D. Tay, H. Pham, Q.C. Pham, Large-scale 3D printing by a team of mobile robots. Automat. Constr. **95**, 98–106 (2018). https://doi.org/10.1016/j.autcon.2018.08.004
98. A.G. Gibb, *Off-Site Fabrication: Prefabrication, Pre-Assembly and Modularisation* (Wiley, 1999) ISBN: 0-470-37836-0
99. J. Wang, S.N. Razavi, Low false alarm rate model for unsafe-proximity detection in construction. J. Comput. Civ. Eng. **30**, 04015005 (2016). https://doi.org/10.1061/(asce)cp.1943-5487.0000470
100. D. Delgado Camacho, P. Clayton, W.J. O'Brien, C. Seepersad, M. Juenger, R. Ferron, S. Salamone, Applications of additive manufacturing in the construction industry – A forward-looking review. Automat. Constr. **89**, 110–119 (2018). https://doi.org/10.1016/j.autcon.2017.12.031
101. E.R. Azar, V.R. Kamat, Earthmoving equipment automation: A review of technical advances and future outlook. ITcon. **22**(2017), 247–265 (2017) https://www.itcon.org/2017/13
102. CPWR – The Center for Construction Research and Training, *Fatal Injuries at Road Construction Sites Among Construction Workers* (CPWR – The Center for Construction Research and Training. https://www.cpwr.com/update_newsletter/fatal-injuries-at-road-construction-sites-among-construction-workers-a-cpwr-quarterly-data-report/. Accessed 11.10.18, 2018)
103. Bureau of Labor Statistics, *Fatal Occupational Injuries by Industry and Event or Exposure* (all United States, 2019) https://www.bls.gov/iif/oshwc/cfoi/cftb0331.htm. Accessed 07.10.19
104. Built Robotics, Robots that build the world 2021. http://www.builtrobotics.com/. Accessed 03.01.21
105. S. Dadhich, U. Bodin, U. Andersson, Key challenges in automation of earth-moving machines. Automat. Constr. **68**, 212–222 (2016). https://doi.org/10.1016/j.autcon.2016.05.009
106. J. Kim, S.S. Lee, J. Seo, V.R. Kamat, Modular data communication methods for a robotic excavator. Automat. Constr. **90**, 166–177 (2018). https://doi.org/10.1016/j.autcon.2018.02.007

107. J.G. Martinez, G. Albeaino, M. Gheisari, W. Volkmann, L.F. Alarcón, UAS point cloud accuracy assessment using structure from motion–based photogrammetry and PPK georeferencing technique for building surveying applications. J. Comput. Civ. Eng. **35**, 05020004 (2021). https://doi.org/10.1061/(asce)cp.1943-5487.0000936
108. F. Elghaish, S. Matarneh, S. Talebi, M. Kagioglou, M.R. Hosseini, S. Abrishami, Toward digitalization in the construction industry with immersive and drones technologies: a critical literature review. Smart Sustain, Built Environ. ahead-of-print (2020). https://doi.org/10.1108/SASBE-06-2020-0077
109. S. Agnisarman, S. Lopes, K. Chalil Madathil, K. Piratla, A. Gramopadhye, A survey of automation-enabled human-in-the-loop systems for infrastructure visual inspection. Automat. Constr. **97**, 52–76 (2018). https://doi.org/10.1016/j.autcon.2018.10.019
110. J.J. Lin, A. Ibrahim, S. Sarwade, M. Golparvar-Fard, Bridge inspection with aerial robots: Automating the entire pipeline of visual data capture, 3D mapping, defect detection, analysis, and reporting. J. Comput. Civ. Eng. **35**, 04020064 (2021). https://doi.org/10.1061/(ASCE)CP.1943-5487.0000954
111. T. Nasiruddin Khilji, L.L.A. Loures, E.R. Azar, Distress recognition in unpaved roads using unmanned aerial systems and deep learning segmentation. J. Comput. Civ. Eng. **35**, 04020061 (2021). https://doi.org/10.1061/(asce)cp.1943-5487.0000952
112. S. Kim, Y. Gan, J. Irizarry, Framework for UAS-integrated airport runway design code compliance using incremental mosaic imagery. J. Comput. Civ. Eng. **35**, 04020070 (2021). https://doi.org/10.1061/(ASCE)CP.1943-5487.0000960
113. D. Liu, X. Xia, J. Chen, S. Li, Integrating building information model and augmented reality for drone-based building inspection. J. Comput. Civ. Eng. **35**, 04020073 (2021). https://doi.org/10.1061/(ASCE)CP.1943-5487.0000958
114. J. Park, K. Kim, Y.K. Cho, Framework of automated construction-safety monitoring using cloud-enabled BIM and BLE mobile tracking sensors. J. Constr. Eng. Manag. **143**, 04020073 (2017). https://doi.org/10.1061/(asce)co.1943-7862.0001223
115. M. Akinlolu, T.C. Haupt, D.J. Edwards, F. Simpeh, A bibliometric review of the status and emerging research trends in construction safety management technologies. Int. J. Constr. Manage. ahead-of-print (2020). https://doi.org/10.1080/15623599.2020.1819584
116. M. Gheisari, A. Rashidi, B. Esmaeili, Using unmanned aerial systems for automated fall hazard monitoring, in *Proceedings of Construction Research Congress 2018*, (American Society of Civil Engineers, New Orleans, LA, 2018), pp. 62–72. https://doi.org/10.1061/9780784481288.007
117. Y. Fang, Y.K. Cho, S. Zhang, E. Perez, Case study of BIM and cloud–enabled real-time RFID indoor localization for construction management applications. J. Constr. Eng. Manag. **142**, 05016003 (2016). https://doi.org/10.1061/(asce)co.1943-7862.0001125
118. M. Wang, C.C. Wang, S. Sepasgozar, S. Zlatanova, A systematic review of digital technology adoption in off-site construction: Current status and future direction towards industry 4.0. Buildings **10**, 204 (2020). https://doi.org/10.3390/buildings10110204
119. J. Louis, P.S. Dunston, Integrating IoT into operational workflows for real-time and automated decision-making in repetitive construction operations. Automat. Constr. **94**, 317–327 (2018). https://doi.org/10.1016/j.autcon.2018.07.005
120. C. Ning, F. You, Data-driven decision making under uncertainty integrating robust optimization with principal component analysis and kernel smoothing methods. Comput. Chem. Eng. **112**, 190–210 (2018). https://doi.org/10.1016/j.compchemeng.2018.02.007
121. V. Kulkarni, R. Sharma, M. Hote, M.E. Civil, Factors affecting material management on construction site. IRJET. **4**, 474–478 (2017) https://www.irjet.net/archives/V4/i1/IRJET-V4I178.pdf
122. N. Kasim, N. Sarpin, H. Mohd Noh, R. Zainal, S. Mohamed, N. Manap, M.Y. Yahya, Automatic materials tracking practices through RFID implementation in construction projects, in *International Conference on Built Environment and Engineering*, (MATEC

Web of Conferences, Johor, Malaysia, 2019), p. 050001. https://doi.org/10.1051/matecconf/201926605001

123. US EPA – United States Environmental Protection Agency, *Advancing Sustainable Materials Management: 2017 Fact Sheet* (United States Environmental Protection Agency, Office of Land and Emergency Management, Washington, DC, 2019), p. 20460. https://www.epa.gov/facts-and-figures-about-materials-waste-and-recycling/advancing-sustainable-materials-management. Accessed 09.09.19
124. M. Pan, T. Linner, W. Pan, H. Cheng, T. Bock, A framework of indicators for assessing construction automation and robotics in the sustainability context. J. Clean. Prod. **182**, 82–95 (2018). https://doi.org/10.1016/j.jclepro.2018.02.053
125. S. Lee, W. Pan, T. Linner, T. Bock, A framework for robot assisted deconstruction: process, sub-systems and modelling, in *Proceedings of the 32nd International Symposium on Automation and Robotics in Construction*, (The International Association for Automation and Robotics in Construction, Oulu, Finland, 2015), pp. 1–8. https://doi.org/10.22260/isarc2015/0093

Chapter 2
Essential Knowledge, Skills, and Abilities Required for Talent Cultivation in Construction Automation and Robotics

Fopefoluwa M. Bademosi and Raja R. A. Issa

Abstract With the global rise in the level of advanced automation and robotics technologies deployed in construction technology and innovation practice, successful implementation of these technologies requires the inclusion of a workforce with the competencies necessary to drive automation and robotics processes. To meet the market demand for these professionals, there is a need to integrate automation and robotics technologies and processes in construction curricula. However, to ensure proper planning of the integration of automation and robotics in construction education, it is necessary to establish a consensus on the competencies required from construction technology and innovation specialists. Using the Delphi method with contributions from professionals, this chapter identifies and examines the fundamental knowledge, skills, and abilities (KSAs) that should be represented in construction education. After a semi-structured interview round and two rounds of the Delphi study, four competency groups with a total of 99 KSA line items were identified, and a consensus was measured for each of the line items with the importance ratings and levels of agreement (LOAs) attained between the professionals. By identifying these essential KSAs, a competency benchmark and performance assessment framework was developed to support industry professionals and educators in talent cultivation in the advancement of construction innovations in automation and robotics.

Keywords Delphi study · Construction · Automation · Robotics · Cloud computing · Big data analytics · Internet of things · Machine learning · Education · Talent cultivation

F. M. Bademosi (✉)
Wentworth Institute of Technology, Boston, MA, USA
e-mail: bademosif@wit.edu

R. R. A. Issa
University of Florida, Gainesville, FL, USA
e-mail: raymond-issa@ufl.edu

H. Jebelli et al. (eds.), *Automation and Robotics in the Architecture, Engineering, and Construction Industry*, https://doi.org/10.1007/978-3-030-77163-8_2

2.1 Introduction

Over the last few years, the construction industry has been faced with a significant skilled labor shortage problem. A study by Autodesk and the Associated General Contractors (AGC) of America revealed that 80% of construction organizations have difficulty finding and hiring skilled craft workers representing the bulk of the construction workforce [1]. The labor gap can be attributed to two underlying trends: an aging workforce retiring quicker than they can be replaced and the lack of interest from younger workers to work in construction [2]. The labor shortage has had a ripple effect on the industry, resulting in cost overruns, schedule delays, and low productivity [1]. Construction companies are addressing the labor shortage by increasing base pay rates and overtime hours and improving the image of the industry to attract younger workers [3].

Furthermore, companies are looking for ways to improve efficiency and productivity through technology [1]. While the industry has typically been slow to adopt new technologies, the growing gap between skilled workers' supply and demand has motivated companies to embrace innovative solutions that automate processes. As automation and robotics technologies progressively become adopted in the architecture, engineering, construction, and operations (AECO) industry, it is expected that there will be an increased demand for construction technology and innovation specialists with the competencies required to successfully perform in the workforce of the industrialized era. However, the limited integration of automation and robotics technologies and processes in construction curricula and the current lack of industry standards and best practices necessitates establishing the fundamental knowledge, skills, and abilities (KSAs) required to develop talent in this domain.

This research is a continuation of formerly published work [4] that builds on the methodology to achieve a consensus on the essential automation, and robotics KSA's construction management (CM) graduates must possess to thrive in the workforce of the futures. By identifying these essential KSAs, a competency benchmark and performance assessment framework was developed to support industry professionals and educators in talent cultivation in automation and robotics.

2.2 Automation and Robotics in Construction Management Education

Automation and robotics technologies present immense benefits for the environment, the economy, and the construction industry on a global scale. Although these advanced technologies are already being implemented in construction, the integration of automation and robotics technologies and processes in construction education curricula is limited.

2.2.1 Current Curriculum Integration

The deployment of automation and robotics technologies in construction are recommended as a solution to enhancing productivity, decreasing the shortages of skilled labor, improving work conditions, and enhancing the quality of the final products in the industry. Construction managers, architects, designers, and engineers have a crucial role to play in the construction and improvement of the built environment. Therefore, it is very important to introduce automation and robotics technologies and processes to construction students in the classroom in order to equip them with the necessary KSAs required to make them successful in their careers in the era of industrialization. Over the years, research and development in the field of automation and robotics has become well established in construction. However, not as much momentum has been achieved in transferring the knowledge gained from research efforts in the classroom [5].

While several researchers have discussed different approaches to curriculum development in automation and robotics in science, technology, engineering, and mathematics (STEM) education, very few have examined this innovative curriculum development in AECO education [6–12]. As automation and robotics technologies continue to develop and achieve noteworthy progress in construction, cultivating and hiring young talent should be of significant importance to the industry. A significant and viable solution to the increasing demand for a strong talent pipeline would be to incorporate automation and robotics in the large AECO curriculum. Faculty in AECO-related programs should evaluate their curriculum to ensure students are exposed to automation and robotics technologies and the inter-disciplinary KSAs required to thrive in the industrialization era.

2.2.2 Addressing the Future of Work in Construction

The workplace of the future is an uncertain place. The world is rapidly changing, and these continuously evolving technologies are changing the fundamental nature of how we learn and work. It is very critical to comprehend the influence automation and robotics technologies could have on contemporary work practices. One major misconception in the era of industrialization and digitization is that automation and robotics technologies will take over our jobs. With these new disruptive technologies on the horizon, it is crucial to understand what jobs will change and predict how these technologies could disrupt current work practices as these innovations bring uncertainty to those whose jobs may be affected by these sudden changes. Data from research house AlphaBeta [13] provides insight into job tasks in specific industries that are susceptible to automation.

For construction managers and supervisors, project managers, and estimators, it forecasts that 13% of the job tasks are more susceptible to automation, while the other 87% have less tendency to be automated [13]. Tasks in these job categories that are inclined to automation include monitoring of facilities or operational systems, collection of data on construction sites, inspections, and monitoring the productivity or efficiency of construction operations. In contrast, 86% of the job tasks of construction laborers are at risk to be automated, with only 14% of their job tasks less predisposed to automation [13].

Other job categories with similar percentages include plasterers, glaziers, floor finishers, tilers, and painting trades workers [13]. By identifying the construction jobs that will be significantly affected by these technologies, organizations can recognize what skillsets to invest in and prepare accordingly. With the skilled labor gap, employees must be properly trained to have the skills required to drive higher levels of construction accuracy as they enter the workforce. Therefore, it is essential to understand and analyze the critical skills that construction employees need to be successful in a speculative, technologically enhanced future.

There is a rising need to develop new skills and mindsets to minimize any risk from this disruption. The digital transformation of the AECO industry calls for a change in the pedagogical approaches in AECO education to adapt to the future of work. In order to equip students adequately with the KSAs needed to succeed in the workforce of the industrialized era, educational institutions must incorporate these emerging technologies and processes into CM education. The future of the AECO industry is highly dependent on the competence of workers entering the workforce.

2.3 Research Methodology

Three phases, literature review, semi-structured interviews, and Delphi study, were used in the implementation of this study to identify the essential KSAs necessary for competency in automation and robotics.

2.3.1 Phase 1: Literature Review

An extensive literature review was completed to establish a foundation for the general KSAs required for research and proficiency in automation and robotics in construction. A content analysis-based review technique was employed to create an initial list of the essential KSAs through a comprehensive academic literature evaluation. The content analysis process of relevant conference proceedings, journals, and academic articles on current state of deployment of automation and robotic technologies in the construction industry, with an emphasis on the United States, is presented in Fig. 2.1.

Content Analysis – Stage 1

- Comprehensive search for literature from 2009 to 2019
- Google Scholar Search Engine
- **Conference proceedings**: ASCE Construction Research Congress (CRC), The Institute of Electrical and Electronics Engineers (IEEE), International Symposium on Automation and Robotics in Construction (ISARC), ASCE Computing Division Computing Workshops, and Joint Conference on Computing in Construction (JC3).
- **Academic journals**: Automation in Construction, Computing in Civil Engineering, Construction Engineering and Management, Information Technology in Construction (ITCon), Educational Technology, Information and Communication Engineering, Management Information Systems, Operations and Production Management, Manufacturing Systems, and Manufacturing Processes
- **Keywords**: automation, robotics technologies, construction, construction education, robotics skills
- 53 related articles selected

Content Analysis – Stage 2

- Screening out irrelevant publications
- 34 publications selected for additional examination
- 29 publications considered

Fig. 2.1 Content analysis-based review method

2.3.2 *Phase 2: Semi-structured Interviews*

To allow further investigation of the research question, semi-structured interviews were conducted on a one-on-one basis, resulting in the collection of more detailed data on the essential KSAs required for automation and robotics proficiency. The expertise and experience of the construction technology and innovation specialists solicited during the interviews were essential in establishing a baseline for categorizing the essential KSAs required for proficiency in automation and robotics. The professionals sought after comprised a range of organizations and were chosen on the basis of their position in the industry, focusing on those who either directly worked with or managed automation and robotics technologies. They comprised of construction technology and innovation specialists, researchers, and educators who have been at the forefront of research, training, and integration of automation and robotics in their organizations.

A participant pool of 32 experts from industry and academia was created, and after initiating contact, 15 committed to participating in the study. The study participants consisted of five VDC professionals, two pre-construction technology managers, two construction technology and innovation managers, three researchers, and three associate professors. The semi-structured interviews allowed for collecting a substantial amount of data, as the interviewees were encouraged to analyze the subject of research and express their views based on their experiences. The findings of the semi-structured interviews were used to corroborate the outcomes of the literature review. Upon completing the interviews, the participants were extended the opportunity to participate in the following Delphi phase of the study.

2.3.3 Phase 3: Delphi Study

The final phase of this study entailed developing and executing a Delphi study with subject matter experts identified during the semi-structured interview phase. The Delphi method is useful in research topics where disagreement or partial knowledge is apparent, and there is a need to achieve consensus [14, 15]. The iterative evaluation procedure aims to allow the Delphi panelists to reconsider their original verdicts on the results emanating from preceding rounds, as the outcomes regarding specific statements can be modified in the later iterations [16]. In theory, the Delphi method can be continuously repeated until an agreement has been achieved. Nonetheless, in most cases, three rounds of iterations are often adequate to obtain the required information to attain a unison [14].

This research phase aimed to achieve a consensus on the essential KSAs expected to meet job-task performance criteria associated with automation and robotics in construction. These KSAs were identified from the literature review and semi-structured interviews, and arriving at a consensus on their importance ratings was crucial in developing a competency benchmark that can be used to cultivate future talent in automation and robotics. Potential participation in the Delphi study was discussed with the participants during the semi-structured interviews, and the expectations about their involvement, time commitment, and expected results of the Delphi study were clearly outlined. The participants who agreed to participate were identified in Phase 2 as experts in the forefront of research, training, and integration of automation and robotics. During the semi-structured interviews, the initial discussion served as an avenue to establish a preliminary list of the required automation and robotics KSAs. A consensus of the KSAs was measured during the Delphi phase, which was conducted in two iterations.

2.3.4 Consensus Measurement in Delphi Studies

While the Delphi method's primary intention is to achieve consensus between the panelists, there is no standard procedure for measuring consensus [15]. To tackle the unclear approach in measuring consensus in Delphi studies, an empirical study was conducted on several research studies from 1975 to 2013 that implemented the Delphi technique [15]. The results of the empirical studies showed that there are predominantly three approaches taken to measure consensus. Some studies used the frequency distribution method to measure agreement, while others used the standard deviation (SD) or the interquartile range (IQR) in measuring consensus. With the frequency distribution method, the percentage of responses to a particular category is defined and determined to be 51%, and there are other situations where a stipulated distance from the mean is also considered.

In the research studies that used the SD or IQR to measure consensus, it is required that the SD be less than 1.5 [17]. For the IQR, there have been contradictory studies where the IQR is required to be less than 2.5 [18] or less than 1.0 [19]. Nevertheless, for every analysis, both the mean and median of the responses must be computed, as they are used in describing the middle and most typical response, showing the central tendency and the coefficient of variation. That is, the division of the SD with the mean denotes the consistency of the observations, and the mode represents the most frequently occurred value. To overcome the disparity in consensus measurement in Delphi studies, a complementary method that employs three measures of assessing consensus concurrently was recommended, as they could not separately be considered a good substitute [15]. These measures are the SD, IQR, and 51% of respondents lying in the "very important" or "important" category.

This recommendation was further explored and developed into several combinative, multicriteria assessment metrics for measuring consensus based on three elements, SD, IQR, and the percent score (PS), shown in Table 2.1 [20]. In this approach, seven distinct levels of agreements (LOAs) were established and allocated to the respondents' results. The first two, early consensus (EC) and early strong agreement (ESA), were determined from the initial responses recorded during the Delphi study's first evaluation phase. The remaining five LOAs, consensus (C), strong agreement (SA), partial agreement (PA), strong disagreement (SD), and total disagreement (TD), were established during the final response and re-evaluation phase of the Delphi [20]. These thorough assessment metrics were projected to assist in the prevention of incorrect measurement of consensus and, to some extent, counteract the possible biases of determining consensus as a result of the reasonably small number of Delphi panelists and the use of a 5-point Likert scale as an alternative to a 10-point scale.

Table 2.1 Combinative multicriteria assessment metrics for LOAs (Adapted from: [20])

LOAs	SD	Operator	IQR Round 1	IQR Round 2	Operator	PS Single	Operator	PS combined adjacent
EC	≤ 1	AND	≤ 1	–	AND	$\geq 60\%$	–	–
ESA	≤ 1	AND	≤ 1	–	–	–	AND	$\geq 70\%$
C	≤ 1	AND	–	≤ 1	AND	$\geq 60\%$	–	–
SA	≤ 1	AND	–	1	–	–	AND	$\geq 70\%$
PA	≤ 1	AND	–	$1 < IQR \leq 2$	–	–	AND	$\geq 60\%$
SD	> 1	OR	–	> 2	AND	$\geq 25\%$	–	–
TD	> 1	OR	–	> 2	OR	$< 60\%$	–	–

EC early consensus, *ESA* early strong agreement, C consensus, *SA* strong agreement, *PA* partial agreement, *SD* strong disagreement, *TD* total disagreement

2.3.5 Survey Development and Study Timeline

The Delphi study was conducted over 2 months, and two rounds of the study were completed to collect opinions and ratings of the automation and robotics KSAs. The Delphi study is aimed to determine LOAs in the responses of the Delphi study panelists and examine the final consensus achieved. The survey questionnaires used in both rounds of the Delphi were developed and managed using Qualtrics survey software. The results from the preceding semi-structured interview phase of the study were incorporated into the first iteration of the Delphi study for review by 15 construction technology and innovation specialists who agreed to participate in the Delphi phase. In both rounds of the Delphi study, responses on the importance ranking of the automation and robotics KSAs were collected from the expert panelists, using a 5-point Likert scale, ranging from "1: Not Important" to "5: Very Important."

Combinative multicriteria evaluation metrics were employed for measuring consensus in the study, as shown in Table 2.1, to establish the LOAs from the resulting importance ratings. The Delphi study's definitive results were collected and reviewed to determine the LOAs achieved for all the automation and robotics KSAs and the consensus reached by the Delphi expert panelists.

2.4 Results

This section elaborates on the results of the different phases of the study. Essential automation and robotics KSAs identified from the literature review are presented and further reviewed during the semi-structured interviews. The data analyzed during the Delphi study phase was used to determine the LOAs in the panel's responses and examine the final consensus achieved on the essential KSAs.

2.4.1 Automation and Robotics KSAs Identified in Research

Detailed results of the literature review have been presented in a previous study [4]. Fundamental KSAs needed to complete the automation and robotics job tasks were extracted from the articles based on a general mention in several applications during the content analysis review process. These defined KSAs were divided into two categories, KSAs for computing and KSAs for computer science, as summarized in Table 2.2.

Table 2.2 Foundational automation and robotics KSAs

KSA groups	KSAs
Computing	BIM
	Cloud computing
	Collaborative environments
	Electronic communications
	Parametric design
	Programming
	Specialized engineering software
Computer science	Algorithms and computational complexity
	Automated reasoning in artificial intelligence
	Big data analytics
	Computational mechanics
	Computer graphics
	Database concepts (computer-based databases)
	Data structures
	Geometric modeling
	Knowledge systems for decision support
	Machine learning
	Object representation and reasoning
	Optimization and search

2.4.2 Review of Essential Automation and Robotics KSAs

The preliminary list of KSAs identified during the literature review was discussed and further expanded with the experts during the semi-structured interviews. After completing the interviews, a total of 99 KSAs were identified and considered for further evaluation during the Delphi phase. Although some of the skills mentioned by the participants during the interviews were termed differently, the connotations were the same. The KSAs were further consolidated into four competency groups: computing, computer science, technology, and attitudes.

2.4.2.1 Computing Competencies

The KSAs under the computing competency group were further divided into four sub-groups: BIM, cloud computing, collaborative environments, and programming. Nine KSAs were assigned to the BIM competency sub-group, six KSAs were assigned to the cloud computing competency sub-group, eight KSAs were assigned to the collaborative environments competency sub-group, and five KSAs were assigned to the programming competency sub-group. The computing competency sub-groups and KSAs are presented in Table 2.3.

Table 2.3 Computing competency sub-groups and KSAs

Competency sub-groups	KSAs
BIM	Understand BIM concepts and how they apply to construction
	Adept in reading and understanding multi-discipline construction plans and specifications
	Ability with multiple BIM applications
	Have knowledge of parametric object-based design
	Perform basic 3D modeling
	Perform advanced modeling, family editing, custom patterns and profiles, and conceptual massing
	Apply BIM-based technology to manage the construction process
	Utilize 3D models for reviewing design and constructability and model coordination
	Perform BIM-based planning and logistics, scheduling, cost estimating, and safety planning
Cloud computing	Have knowledge of cloud-based BIM platforms
	Use cloud-based systems to access project data
	Employ cloud-based platforms for BIM visualization and for efficient acquisition of applicable information through the web
	Apply cloud computing for construction collaboration, coordination, and communication
	Store building data during design and construction and over the entire life of the building on the cloud
	Use cloud platforms for information exchange and mobile access to information
Collaborative environments	Understand technology for collaborative systems
	Ability to work in a computer-supported cooperative environment
	Define clear outline of responsibilities
	Develop and adhere to common processes and tools
	Work efficiently within co-located and remote teams and foster virtual teamwork
	Develop trust among team members
	Establish and communicate conflict resolution strategies
	Willingness to share resources among project participants
Programming	Have computer science knowledge
	Comprehend core programming languages
	Ability to write codes
	Perform computational design using dynamo
	Create applications or plug-ins

2.4.2.2 Computer Science Competencies

The computer science competency group was further broken down into four sub-groups: big data analytics, database concepts, Internet of things, and machine learning. Five KSAs were assigned to the big data analytics competency sub-group, four KSAs were assigned to the database concepts competency sub-group, five KSAs were assigned to the Internet of things competency sub-group, and six KSAs were assigned to the machine learning competency sub-group. The computer science competency sub-groups and KSAs are shown in Table 2.4.

Table 2.4 Computer science competency sub-groups and KSAs

Competency sub-groups	KSAs
Big data analytics	Ability to visualize data and communicate
	Collect, analyze, and compare project data for reporting and to determine project feasibility
	Develop project benchmarks that define project success
	Collect data for project benchmarks and report as necessary to project stakeholders
	Identify likelihood of future outcomes based on past project data
Database concepts	Have knowledge about computer-based databases
	Examine large databases to get actionable insights
	Extract concepts, interrelationships, and patterns of interest from large databases
	Analyze large amount of construction project data to identify valid, useful, and previously unknown patterns
Internet of things	Identify use cases for IoT in construction
	Export BIM content to various formats for use in IoT applications
	Install sensors on physical devices and program sensors and connect to an online system
	Monitor sensors and devices to produce and communicate reports based on sensor data and feedback
	Ability to store data securely
Machine learning	Understand concepts of machine learning and artificial intelligence
	Have knowledge of computer science fundamentals – Data structures, algorithms, computability and complexity, computer architecture
	Perform data modeling and evaluation
	Apply probability and statistics to learn from data and provide insights to build intelligent applications
	Apply machine learning algorithms and libraries
	Configure robotic process automation to automate workflows and perform tasks

2.4.2.3 Technology Competencies

The technology competency group was further divided into three sub-groups: reality computing, visualization, and automation. Seven KSAs were assigned to the reality computing competency sub-group, five KSAs were assigned to the visualization competency sub-group, and nine KSAs were assigned to the automation competency sub-group. The technology competency sub-groups and KSAs are presented in Table 2.5.

Table 2.5 Technology competency sub-groups and KSAs

Competency sub-groups	KSAs
Reality computing	Use reality capture tools for construction applications
	Understand scopes of work and how reality capture may benefit a project
	Use reality capture tools to collect as-built conditions
	Integrate 3D models with reality capture information
	Ensure point clouds and 3D models are coordinated and accurate
	Use reality capture software to extract datasets from images or scans
	Understand drone use, insurance requirements, FAA regulations
Visualization	Understand visualization technologies and their applications
	Develop immersive/interactive visualizations and interact in them
	Import 3D models into immersive/interactive environments
	Use visualization for design and constructability review
	Perform renders and visualizations of 3D models
Automation	Have knowledge of automation and robotics trends and applications in construction
	Identify use cases for using robotics in construction
	Understand the basic principle of robotic and automation systems
	Develop 3D models with supplemental detailing to create building production models
	Export 3D model content to various formats for use in 3D printing, manufacturing, prefabrication, etc.
	Monitor and update models and drawings based on feedback from production
	Program and manipulate robotic systems for specific tasks
	Use iterative programs to create generative designs that optimize construction
	Use artificial intelligence to optimize construction schedule, cost, material cuts, and quantities

2.4.2.4 Attitudes Competencies

The KSAs under the attitudes competency group were further divided into six sub-groups: technology confidence, active learning, judgment and decision-making, complex problem-solving, innovation and creativity, and leadership. Four KSAs were assigned to the technology confidence competency sub-group, five KSAs were assigned to the active learning competency sub-group, six KSAs were assigned to the judgment and decision-making competency sub-group, three KSAs were assigned to the complex problem-solving competency sub-group, five KSAs were assigned to the innovation and creativity competency sub-group, and seven KSAs were assigned to the leadership competency sub-group. The attitudes competency sub-groups and KSAs are shown in Table 2.6.

2.4.3 Consensus Building of Automation and Robotics KSAs

After identifying ninety-nine (99) automation and robotics KSAs during the semi-structured interviews, the KSAs were further evaluated during the Delphi study phase, to determine consensus on the rating and ranking of the essential KSAs required for talent cultivation in automation and robotics.

2.4.3.1 Round 1: Initial Evaluation and Rating

The purpose of Round 1 of the Delphi study was to determine the LOAs on the importance of the KSAs necessary to meet job-task performance criteria associated with automation and robotics based on the Delphi panel's responses to the 5-point Likert scale. At the end of Round 1, 15 completed questionnaires were received, resulting in a response rate of 100%. The importance ratings of each robotics KSA were collated, and the SD, IQR, and PS were determined as a result. The aim of Round 1 of the study was to identify whether any of the KSAs achieved levels of either EC or ESA. This was verified by assessing the resulting SD, IQR, and PS when compared to the combinative multicriteria assessment metrics.

The SD, IQR, and PS of the importance ratings of the KSAs were calculated and measured for consensus utilizing the Round 1 combinative multicriteria assessment metrics shown in Fig. 2.2. EC represents the highest LOA per definition, where 60% or more of the Delphi panel rated the KSA as very important. ESA is the second-highest LOA defined, where the combined adjacent scale of the responses of the Delphi panelists (either 4 and 5 or 1 and 2) PS was $\geq$70%. TBD depicts the KSAs subject to further LOA evaluation, which is the group of KSAs to be carried forward into Round 2 for re-rating by the Delphi panel. The results of the evaluation process in Round 1, showing the KSAs that achieved LOAs of EC and ESAs, are shown in Figs. 2.3, 2.4, 2.5 and 2.6.

Table 2.6 Attitudes competency sub-groups and KSAs

Competency sub-groups	KSAs
Technology confidence	Familiar and current with emerging technologies in construction
	Technology-savvy
	Enthusiastic about the potentials of new technologies
	Generate or adapt technology to serve the company's needs
Active learning	Motivation to continually learn
	Learn new software and hardware as necessary
	Ability to adapt knowledge to continuous updates and changes of technologies
	Participate in industry events to stay up to date with emergent technologies
	Ability to stay updated and grasp new concepts quickly
Judgment and decision-making	Define problems, challenges, or opportunities for new technologies
	Generate an array of possible solutions or applications
	Evaluate the costs and benefits, or pros and cons, associated with each technology
	Select the optimum solution or technology option
	Implement the solution/option chosen
	Assess the impact of the decision and modify the course of action as needed
Complex problem-solving	Identify complex problems inherent in the construction process and review-related data
	Develop and evaluate alternatives and implement solutions
	Select the best solution and technology befitting the problem
Innovation and creativity	Be innately curious and creative
	Discover novel ways to approach specific problems
	Ability to generate and try out innovative ideas
	Be keen to make technical innovations
	Develop new ideas for and answers to work-related problems through creative and alternative thinking
Leadership	Identify the training needs of the project team
	Participate in or present training
	Develop training programs or classes and training materials
	Be willing to lead, take charge, and offer opinions and direction regarding technology integration
	Define new workflows using new technology
	Present new technology workflows internally
	Participate in pilot programs utilizing new technology

The summary of the results of the consensus evaluation in Round 1 are shown in Fig. 2.7 and are as follows:

1. 28 KSAs in the computing competency group were analyzed. Out of the 28 KSAs, 25% achieved a LOA of EC, and 46% achieved a LOA of ESA, while the LOAs of the remaining 29% are yet to be determined.

ROUND 1 - DELPHI STUDY LEVELS OF AGREEMENT (LOAs)								
Level of Agreement		Combinatory Multicriteria Evaluation Metrics for Levels of Agreement						
		SD	Operator	IQR	Operator	PS (Single)	Operator	PS (Combined Adjacent)
EC	Early Consensus	≤ 1	AND	≤ 1	AND	≥ 60%	—	—
ESA	Early Strong	≤ 1	AND	≤ 1	—	—	AND	≥ 70%
TBD	To Be Determined	> 1	OR	> 1	—	—	AND	< 70%

Fig. 2.2 Round 1 combinative multicriteria evaluation metrics

ROUND 1 - DELPHI ASSESSMENT OF CONSTRUCTION ROBOTICS TECHNOLOGIES KSAs																				
KNOWLEDGE, SKILLS AND ABILITIES			MIN	MAX	MEAN	MEDIAN	MODE	VAR	STD	Q1	Q3	IQR	% = 5	% = 4	% = 3	% = 2	% = 1	% 4,5	% 1,2	LOA (R1)
COMPUTING COMPETENCY GROUP	Building Information Modeling	Understand BIM concepts and how they apply to construction	4	5	4.80	5.00	5.00	0.17	0.41	5	5	0	80.00%	20.00%	0.00%	0.00%	0.00%	100%	0%	EC
		Adept in reading and understanding multi-discipline construction plans and specifications	2	5	4.47	5.00	5.00	0.70	0.83	4	5	1	60.00%	33.33%	0.00%	6.67%	0.00%	93%	7%	EC
		Ability with multiple BIM applications	3	5	4.40	4.00	4.00	0.40	0.63	4	5	1	46.67%	46.67%	6.67%	0.00%	0.00%	93%	0%	ESA
		Have knowledge of parametric object-based design	1	5	3.93	4.00	5.00	1.50	1.22	3.5	5	1.5	40.00%	33.33%	13.33%	6.67%	6.67%	73%	13%	TBD
		Perform basic 3D modeling	3	5	4.40	5.00	5.00	0.54	0.74	4	5	1	53.33%	33.33%	13.33%	0.00%	0.00%	87%	0%	ESA
		Perform advanced modeling, family editing, custom patterns and profiles, and conceptual massing	2	5	4.07	4.00	5.00	1.07	1.03	4	5	1	40.00%	40.00%	6.67%	13.33%	0.00%	80%	13%	TBD
		Apply BIM-based technology to manage the construction process	4	5	4.53	5.00	5.00	0.27	0.52	4	5	1	53.33%	46.67%	0.00%	0.00%	0.00%	100%	0%	ESA
		Utilize 3D models for reviewing design and constructability, and model coordination	3	5	4.60	5.00	5.00	0.40	0.63	4	5	1	66.67%	26.67%	6.67%	0.00%	0.00%	93%	0%	EC
		Perform BIM-based planning and logistics, scheduling, cost estimating, and safety planning	3	5	4.40	5.00	5.00	0.69	0.83	4	5	1	60.00%	20.00%	20.00%	0.00%	0.00%	80%	0%	EC
	Cloud Computing	Have knowledge of cloud-based BIM platforms	3	5	4.40	5.00	5.00	0.54	0.74	4	5	1	53.33%	33.33%	13.33%	0.00%	0.00%	87%	0%	ESA
		Use cloud-based systems to access prroject data	3	5	4.53	5.00	5.00	0.41	0.64	4	5	1	60.00%	33.33%	6.67%	0.00%	0.00%	93%	0%	EC
		Employ cloud-based platforms for BIM visualization and for efficient acquisition of applicable information through the web	3	5	4.53	5.00	5.00	0.41	0.64	4	5	1	60.00%	33.33%	6.67%	0.00%	0.00%	93%	0%	EC
		Apply cloud computing for construction collaboration, coordination, and communication	3	5	4.40	4.00	4.00	0.40	0.63	4	5	1	46.67%	46.67%	6.67%	0.00%	0.00%	93%	0%	ESA
		Store building data during design and construction and over the entire life of the building on the cloud	2	5	4.20	4.00	4.00	0.74	0.86	4	5	1	40.00%	46.67%	6.67%	6.67%	0.00%	87%	7%	ESA
		Use cloud platforms for Information exchange and mobile access to information	2	5	4.53	5.00	5.00	0.70	0.83	4	5	1	66.67%	26.67%	0.00%	6.67%	0.00%	93%	7%	EC
	Collaborative Environments	Understand technology for collaborative systems	2	5	4.33	4.00	4.00	0.67	0.82	4	5	1	46.67%	46.67%	0.00%	6.67%	0.00%	93%	7%	ESA
		Ability to work in a computer-supported cooperative environment	2	5	4.20	4.00	4.00	0.74	0.86	4	5	1	40.00%	46.67%	6.67%	6.67%	0.00%	87%	7%	ESA
		Define clear outline of responsibilities	1	5	3.80	4.00	4.00	1.46	1.21	4	4	0	20.00%	66.67%	0.00%	0.00%	13.33%	87%	13%	TBD
		Develop and adhere to common processes and tools	2	5	4.20	4.00	4.00	0.74	0.86	4	5	1	33.33%	53.33%	6.67%	6.67%	0.00%	87%	7%	ESA
		Work efficiently within co-located and remote teams and foster virtual teamwork	3	5	4.47	5.00	5.00	0.41	0.64	4	5	1	53.33%	40.00%	6.67%	0.00%	0.00%	93%	0%	ESA
		Develop trust among team members	3	5	4.20	4.00	4.00	0.60	0.77	4	5	1	40.00%	40.00%	20.00%	0.00%	0.00%	80%	0%	ESA
		Establish and communicate conflict resolution strategies	1	5	3.93	4.00	5.00	1.78	1.33	3.5	5	1.5	46.67%	26.67%	6.67%	13.33%	6.67%	73%	20%	TBD
		Willingness to share resources among project participants	1	5	3.73	4.00	5.00	2.21	1.49	3	5	2	40.00%	33.33%	0.00%	13.33%	13.33%	73%	27%	TBD
	Programming	Have computer science knowledge	2	5	4.20	4.00	5.00	0.89	0.94	4	5	1	46.67%	33.33%	13.33%	6.67%	0.00%	80%	7%	ESA
		Comprehend core programming languages	1	5	4.00	4.00	5.00	1.43	1.20	3	5	2	46.67%	20.00%	26.67%	0.00%	6.67%	67%	7%	TBD
		Ability to write codes	2	5	3.73	4.00	4.00	0.78	0.88	3.5	4	0.5	13.33%	60.00%	13.33%	13.33%	0.00%	73%	13%	ESA
		Perform computational design using Dynamo	1	5	3.60	4.00	4.00	1.54	1.24	3	4.5	1.5	26.67%	33.33%	20.00%	13.33%	6.67%	60%	20%	TBD
		Create applications or plug-ins	1	5	3.00	4.00	4.00	1.86	1.36	2	4	2	6.67%	46.67%	6.67%	20.00%	20.00%	53%	40%	TBD

Fig. 2.3 Round 1 consensus evaluation results

ROUND 1 - DELPHI ASSESSMENT OF CONSTRUCTION ROBOTICS TECHNOLOGIES KSAs																				
KNOWLEDGE, SKILLS AND ABILITIES			MIN	MAX	MEAN	MEDIAN	MODE	VAR	STD	Q1	Q3	IQR	% = 5	% = 4	% = 3	% = 2	% = 1	% 4,5	% 1,2	LOA (R1)
COMPUTER SCIENCE COMPETENCY GROUP	Big Data Analytics	Ability to visualize data and communicate	4	5	4.60	5.00	5.00	0.26	0.51	4	5	1	60.00%	40.00%	0.00%	0.00%	0.00%	100%	0%	EC
		Collect, analyze and compare project data for reporting and to determine project feasibility	2	5	4.07	4.00	4.00	0.50	0.70	4	4	0	20.00%	73.33%	0.00%	6.67%	0.00%	93%	7%	ESA
		Develop project benchmarks that define project success	2	5	3.80	4.00	4.00	1.03	1.01	3	4.5	1.5	26.67%	40.00%	20.00%	13.33%	0.00%	67%	13%	TBD
		Collect data for project benchmarks and report as necessary to project stakeholders	2	5	3.73	4.00	4.00	1.07	1.03	3.5	4	0.5	20.00%	53.33%	6.67%	20.00%	0.00%	73%	20%	TBD
		Identify likelihood of future outcomes based on past project data	2	5	3.93	4.00	4.00	1.07	1.03	3.5	5	1.5	33.33%	40.00%	13.33%	13.33%	0.00%	73%	13%	TBD
	Database Concepts	Have knowledge about computer-based databases	2	5	3.93	4.00	4.00	0.78	0.88	3.5	4.5	1	26.67%	46.67%	20.00%	6.67%	0.00%	73%	7%	ESA
		Examine large databases to get actionable insights	2	5	3.67	4.00	4.00	1.38	1.18	2.5	4.5	2	26.67%	40.00%	6.67%	26.67%	0.00%	67%	27%	TBD
		Extract concepts, interrelationships, and patterns of interest from large databases	1	5	3.60	4.00	4.00	1.54	1.24	3	4.5	1.5	26.67%	33.33%	20.00%	13.33%	6.67%	60%	20%	TBD
		Analyze large amount of construction project data to identify valid, useful, and previously unknown patterns	1	5	3.67	4.00	5.00	1.95	1.40	2.5	5	2.5	36.36%	9.09%	18.18%	27.27%	9.09%	45%	36%	TBD
	Internet of Things	Identify use cases for IoT in construction	2	5	4.27	5.00	5.00	0.92	0.96	4	5	1	53.33%	26.67%	13.33%	6.67%	0.00%	80%	7%	ESA
		Export BIM content to various formats for use in IoT applications	2	5	3.80	4.00	4.00	0.89	0.94	3.5	4	0.5	20.00%	53.33%	13.33%	13.33%	0.00%	73%	13%	ESA
		Install sensors on physical devices, and program sensors and connect to an online system	1	5	3.67	4.00	4.00	1.24	1.11	3	4	1	20.00%	46.67%	20.00%	6.67%	6.67%	67%	13%	TBD
		Monitor sensors and devices to produce and communicate reports based on sensor data and feedback	1	5	3.73	4.00	4.00	1.35	1.16	3	4.5	1.5	26.67%	40.00%	20.00%	6.67%	6.67%	67%	13%	TBD
		Ability to store data securely	2	5	3.80	4.00	4.00	0.89	0.94	3.5	4	0.5	20.00%	53.33%	13.33%	13.33%	0.00%	73%	13%	ESA
	Machine Learning	Understand concepts of machine learning and artificial intelligence	2	5	3.80	4.00	4.00	0.74	0.86	3	4	1	20.00%	46.67%	26.67%	6.67%	0.00%	67%	7%	TBD
		Have knowledge of computer science fundamentals - data structures, algorithms, computability and complexity, computer architecture	1	4	3.20	4.00	4.00	1.31	1.15	2.5	4	1.5	0.00%	60.00%	13.33%	13.33%	13.33%	60%	27%	TBD
		Perform data modeling and evaluation	2	5	3.47	4.00	4.00	1.27	1.13	2	4	2	13.33%	53.33%	0.00%	33.33%	0.00%	67%	33%	TBD
		Apply probability and statistics to learn from data and provide insights to build intelligent applications	1	5	3.40	4.00	4.00	1.54	1.24	3	4	1	6.67%	66.67%	0.00%	13.33%	13.33%	73%	27%	TBD
		Apply machine learning algorithms and libraries	2	5	3.60	4.00	4.00	1.11	1.06	3	4	1	13.33%	60.00%	0.00%	26.67%	0.00%	73%	27%	TBD
		Configure robotic process automation to automate workflows and perform tasks	1	5	3.53	4.00	4.00	1.70	1.30	3	4	1	20.00%	46.67%	13.33%	6.67%	13.33%	67%	20%	TBD

Fig. 2.4 Round 2 combinative multicriteria evaluation metrics

ROUND 1 - DELPHI ASSESSMENT OF CONSTRUCTION ROBOTICS TECHNOLOGIES KSAs																				
KNOWLEDGE, SKILLS AND ABILITIES			MIN	MAX	MEAN	MEDIAN	MODE	VAR	STD	Q1	Q3	IQR	% = 5	% = 4	% = 3	% = 2	% = 1	% 4,5	% 1,2	LOA (R1)
TECHNOLOGY COMPETENCY GROUP	Reality Computing	Use reality capture tools for construction applications	4	5	4.40	4.00	4.00	0.26	0.51	4	5	1	40.00%	60.00%	0.00%	0.00%	0.00%	100%	0%	ESA
		Understand scopes of work and how reality capture may benefit the project	3	5	4.40	4.00	4.00	0.40	0.63	4	5	1	46.67%	46.67%	6.67%	0.00%	0.00%	93%	0%	ESA
		Use reality capture tools to collect as-built conditions	4	5	4.60	5.00	5.00	0.26	0.51	4	5	1	60.00%	40.00%	0.00%	0.00%	0.00%	100%	0%	EC
		Integrate 3D models with reality capture information	3	5	4.60	5.00	5.00	0.40	0.63	4	5	1	66.67%	26.67%	6.67%	0.00%	0.00%	93%	0%	EC
		Ensure point clouds and 3D models are coordinated and accurate	4	5	4.47	4.00	4.00	0.27	0.52	4	5	1	46.67%	53.33%	0.00%	0.00%	0.00%	100%	0%	ESA
		Use reality capture software to extract datasets from images or laser scans	4	5	4.53	5.00	5.00	0.27	0.52	4	5	1	53.33%	46.67%	0.00%	0.00%	0.00%	100%	0%	ESA
		Understand drone use, insurance requirements, FAA regulations	3	5	4.40	5.00	5.00	0.54	0.74	4	5	1	53.33%	33.33%	13.33%	0.00%	0.00%	87%	0%	ESA
	Visualization	Understand visualization technologies and their applications in construction	4	5	4.53	5.00	5.00	0.27	0.52	4	5	1	53.33%	46.67%	0.00%	0.00%	0.00%	100%	0%	ESA
		Develop immersive/interactive visualization environments and interact in them	3	5	4.40	4.00	4.00	0.40	0.63	4	5	1	46.67%	46.67%	6.67%	0.00%	0.00%	93%	0%	ESA
		Import 3D models into immersive/interactive environments	3	5	4.40	4.00	4.00	0.40	0.63	4	5	1	46.67%	46.67%	6.67%	0.00%	0.00%	93%	0%	ESA
		Use visualization technologies for design and constructability review	4	5	4.47	4.00	4.00	0.27	0.52	4	5	1	46.67%	53.33%	0.00%	0.00%	0.00%	100%	0%	ESA
		Perform renders and visualizations of 3D models	3	5	4.47	5.00	5.00	0.55	0.74	4	5	1	60.00%	26.67%	13.33%	0.00%	0.00%	87%	0%	EC
	Automation	Have knowledge of robotics and automation trends and applications in construction	4	5	4.60	5.00	5.00	0.26	0.51	4	5	1	60.00%	40.00%	0.00%	0.00%	0.00%	100%	0%	EC
		Identify use cases for using robotics in construction	4	5	4.67	5.00	5.00	0.24	0.49	4	5	1	66.67%	33.33%	0.00%	0.00%	0.00%	100%	0%	EC
		Understand the basic principle of robotic and automation systems	3	5	4.33	4.00	4.00	0.38	0.62	4	5	1	40.00%	53.33%	6.67%	0.00%	0.00%	93%	0%	ESA
		Develop 3D models with supplemental detailing to create building production models	3	5	4.33	4.00	4.00	0.38	0.62	4	5	1	40.00%	53.33%	6.67%	0.00%	0.00%	93%	0%	ESA
		Export 3D model content to various formats for use in 3D printing, manufacturing, prefabrication, etc.	2	5	3.93	4.00	4.00	0.78	0.88	3.5	4.5	1	26.67%	46.67%	20.00%	6.67%	0.00%	73%	7%	ESA
		Monitor and update models and drawings based on feedback from production	2	5	4.07	4.00	4.00	1.07	1.03	4	5	1	40.00%	40.00%	6.67%	13.33%	0.00%	80%	13%	TBD
		Program and manipulate robotic systems for specific construction tasks	2	5	3.80	4.00	4.00	1.03	1.01	3	4.5	1.5	26.67%	40.00%	20.00%	13.33%	0.00%	67%	13%	TBD
		Use iterative programs to create generative designs that optimize construction	2	5	3.93	4.00	4.00	1.07	1.03	3.5	5	1.5	33.33%	40.00%	13.33%	13.33%	0.00%	73%	13%	TBD
		Use artificial intelligence to optimize construction schedule, cost, material outs and quantities	2	5	4.07	4.00	5.00	1.21	1.10	3.5	5	1.5	46.67%	26.67%	13.33%	13.33%	0.00%	73%	13%	TBD

Fig. 2.5 Round 2 and final consensus evaluation results

ROUND 1 - DELPHI ASSESSMENT OF CONSTRUCTION ROBOTICS TECHNOLOGIES KSAs																				
KNOWLEDGE, SKILLS AND ABILITIES			MIN	MAX	MEAN	MEDIAN	MODE	VAR	STD	Q1	Q3	IQR	% = 5	% = 4	% = 3	% = 2	% = 1	% 4,5	% 1,2	LOA (R1)
ATTITUDES COMPETENCY GROUP	Technology Confidence	Familiar and up-to-date with emerging technologies in construction	3	5	4.47	5.00	5.00	0.41	0.64	4	5	1	53.33%	40.00%	6.67%	0.00%	0.00%	93%	0%	ESA
		Technology-savvy	3	5	4.60	5.00	5.00	0.40	0.63	4	5	1	66.67%	26.67%	6.67%	0.00%	0.00%	93%	0%	EC
		Enthusiastic about the potentials of new technologies	3	5	4.53	5.00	5.00	0.41	0.64	4	5	1	60.00%	33.33%	6.67%	0.00%	0.00%	93%	0%	EC
		Generate or adapt technology to serve the company's needs	3	5	4.47	5.00	5.00	0.41	0.64	4	5	1	53.33%	40.00%	6.67%	0.00%	0.00%	93%	0%	ESA
	Active Learning	Motivation to continually learn	4	5	4.67	5.00	5.00	0.24	0.49	4	5	1	66.67%	33.33%	0.00%	0.00%	0.00%	100%	0%	EC
		Learn new software and hardware as necessary	4	5	4.60	5.00	5.00	0.26	0.51	4	5	1	60.00%	40.00%	0.00%	0.00%	0.00%	100%	0%	EC
		Ability to adapt knowledge to continuous updates and changes of technologies	4	5	4.53	5.00	5.00	0.27	0.52	4	5	1	53.33%	46.67%	0.00%	0.00%	0.00%	100%	0%	ESA
		Participate in industry events to stay up-to-date with emergent technologies	4	5	4.40	4.00	4.00	0.26	0.51	4	5	1	40.00%	60.00%	0.00%	0.00%	0.00%	100%	0%	ESA
		Ability to stay updated and grasp new concepts quickly	4	5	4.67	5.00	5.00	0.24	0.49	4	5	1	66.67%	33.33%	0.00%	0.00%	0.00%	100%	0%	EC
	Judgement and Decision Making	Define problems, challenges, or opportunities for new technologies	4	5	4.67	5.00	5.00	0.24	0.49	4	5	1	66.67%	33.33%	0.00%	0.00%	0.00%	100%	0%	EC
		Generate an array of possible solutions or applications	4	5	4.40	4.00	4.00	0.26	0.51	4	5	1	40.00%	60.00%	0.00%	0.00%	0.00%	100%	0%	ESA
		Evaluate the costs and benefits, or pros and cons, associated with each technology	4	5	4.47	4.00	4.00	0.27	0.52	4	5	1	46.67%	53.33%	0.00%	0.00%	0.00%	100%	0%	ESA
		Select the optimum solution or technology option	4	5	4.47	4.00	4.00	0.27	0.52	4	5	1	46.67%	53.33%	0.00%	0.00%	0.00%	100%	0%	ESA
		Implement the solution/option chosen	4	5	4.80	5.00	5.00	0.17	0.41	5	5	0	80.00%	20.00%	0.00%	0.00%	0.00%	100%	0%	EC
		Assess the impact of the decision and modify the course of action as needed	2	5	4.47	5.00	5.00	0.70	0.83	4	5	1	62.50%	31.25%	0.00%	6.25%	0.00%	94%	6%	EC
	Complex Problem Solving	Identify complex problems inherent in the construction process and review related data	4	5	4.47	4.00	4.00	0.27	0.52	4	5	1	46.67%	53.33%	0.00%	0.00%	0.00%	100%	0%	ESA
		Develop and evaluate alternatives and implement solutions	4	5	4.53	5.00	5.00	0.27	0.52	4	5	1	53.33%	46.67%	0.00%	0.00%	0.00%	100%	0%	ESA
		Select the best solution and technology befitting the problem	4	5	4.60	5.00	5.00	0.26	0.51	4	5	1	60.00%	40.00%	0.00%	0.00%	0.00%	100%	0%	EC
	Innovation and Creativity	Be innately curious and creative	4	5	4.67	5.00	5.00	0.24	0.49	4	5	1	66.67%	33.33%	0.00%	0.00%	0.00%	100%	0%	EC
		Discover novel ways to approach specific problems	4	5	4.60	5.00	5.00	0.26	0.51	4	5	1	60.00%	40.00%	0.00%	0.00%	0.00%	100%	0%	EC
		Ability to generate and try out innovative ideas	4	5	4.73	5.00	5.00	0.21	0.46	4.5	5	0.5	73.33%	26.67%	0.00%	0.00%	0.00%	100%	0%	EC
		Be keen to make technical innovations	3	5	4.53	5.00	5.00	0.41	0.64	4	5	1	60.00%	33.33%	6.67%	0.00%	0.00%	93%	0%	EC
		Develop new ideas for and answers to work-related problems through creative and alternative thinking	4	5	4.53	5.00	5.00	0.27	0.52	4	5	1	53.33%	46.67%	0.00%	0.00%	0.00%	100%	0%	ESA
	Leadership	Identify the training needs of the project team	4	5	4.60	5.00	5.00	0.26	0.51	4	5	1	60.00%	40.00%	0.00%	0.00%	0.00%	100%	0%	EC
		Participate in or present training	4	5	4.47	4.00	4.00	0.27	0.52	4	5	1	46.67%	53.33%	0.00%	0.00%	0.00%	100%	0%	ESA
		Develop training programs or classes, and training materials	3	5	4.47	5.00	5.00	0.41	0.64	4	5	1	53.33%	40.00%	6.67%	0.00%	0.00%	93%	0%	ESA
		Be willing to lead, take charge, and offer opinions and direction regarding technology integration	4	5	4.67	5.00	5.00	0.24	0.49	4	5	1	66.67%	33.33%	0.00%	0.00%	0.00%	100%	0%	EC
		Define new workflows using new technology	4	5	4.40	4.00	4.00	0.26	0.51	4	5	1	40.00%	60.00%	0.00%	0.00%	0.00%	100%	0%	ESA
		Present new technology workflows internally	4	5	4.40	4.00	4.00	0.26	0.51	4	5	1	40.00%	60.00%	0.00%	0.00%	0.00%	100%	0%	ESA
		Participate in pilot programs utilizing new technology	2	5	4.40	5.00	5.00	0.69	0.83	4	5	1	53.33%	40.00%	0.00%	6.67%	0.00%	93%	7%	ESA

Fig. 2.6 Percentages of calculated LOAs during Round 1

ROUND 1 - DELPHI ASSESSMENT OF CONSTRUCTION ROBOTICS TECHNOLOGIES KSAs				
Competency Group	KSA Line Items	Percentage of EC Achieved	Percentage of ESA Achieved	Percentage of TBD Achieved
Computing	28	25%	46%	29%
Computer Science	20	5%	25%	70%
Technology	21	24%	57%	19%
Attitudes	30	50%	50%	0%

Fig. 2.7 Percentages of final LOAs during Round 2

2. 20 KSAs in the computer science competency group were analyzed. Out of the 20 KSAs, 5% achieved a LOA of EC, and 25% achieved a LOA of ESA, while the LOAs of the remaining 70% are yet to be determined.
3. 21 KSAs in the technology competency group were analyzed. Out of the 21 KSAs, 24% achieved a LOA of EC, and 57% achieved a LOA of ESA, while the LOAs of the remaining 19% are yet to be determined.
4. 30 KSAs in the attitudes competency group were analyzed, all of which achieved either EC or ESA LOA. Fifty percent achieved a LOA of EC, and 50% achieved a LOA of ESA.

2.4.3.2 Round 2: Final Evaluation and Rating

Round 2 of the Delphi study was conducted over a 1-month period and allowed the expert panelists to review the cumulative results, assess consensus levels, provide comments, and re-rate the automation and robotics KSAs. At the commencement of Round 2, the expert panelists were given a report of the results from Round 1 and informed about the exclusion of the KSAs that attained both EC and ESA LOAs. Subsequently, the panelists were asked to re-rate the outstanding KSAs to reach consensus by taking into consideration the responses of the other panelists in Round 1. The 15 construction technology and innovation specialists who committed to participating in the Delphi study completed the Round 2 Delphi survey resulting in a 100% response rate.

All automation and robotics KSAs that attained a LOA of either EC or ESA in Round 1 were exempted from Round 2 of the Delphi study. The rest of the KSAs that temporarily achieved a LOA level of TBD were carried forward into Round 2 for further evaluation and re-rating by the Delphi panelists. With all the KSAs in the attitudes competency group achieving LOAs of EC or ESA in Round 1, the evaluation process of the automation and robotics KSAs in Round 2 was divided into three categories. The SD, IQR, and PS of the responses to the outstanding automation and robotics KSAs were calculated and re-evaluated for consensus measurement by utilizing the Round 2 combinative multicriteria assessment metrics and additional defined LOAs as shown in Fig. 2.8.

With Round 2 being the final phase of the Delphi study, the concluding LOAs attained on all the automation and robotics KSAs were assigned in accordance with

ROUND 2 - DELPHI STUDY LEVELS OF AGREEMENT (LOAs)								
Level of Agreement		Combinatory Multicriteria Evaluation Metrics for Levels of Agreement						
		SD	Operator	IQR	Operator	PS (Single)	Operator	PS (Combined Adjacent)
C	Consensus	≤ 1	AND	≤ 1	AND	≥ 60%	-	-
SA	Strong Agreement	≤ 1	AND	1	-		AND	≥ 70%
PA	Partial Agreement	≤ 1	AND	1<IQR≤2	-		AND	≥ 60%
SD	Strong Disagreement	> 1	OR	> 2	AND	25%	-	-
TD	Total Disagreement	> 1	OR	> 2	OR	< 60%	-	-

Fig. 2.8 Excerpt of the framework showing the three levels of training

ROUND 2 - DELPHI ASSESSMENT OF CONSTRUCTION ROBOTICS TECHNOLOGIES KSAs																					
		KNOWLEDGE, SKILLS AND ABILITIES	MIN	MAX	MEAN	MEDIAN	MODE	VAR	STD	Q1	Q3	IQR	% = 5	% = 4	% = 3	% = 2	% = 1	% 4,5	% 1,2	LOA (R1)	LOA (R2)
COMPUTING COMPETENCY GROUP	Building Information Modeling	Understand BIM concepts and how they apply to construction	4	5	4.80	5.00	5.00	0.17	0.41	5	5	0	80.00%	20.00%	0.00%	0.00%	0.00%	100%	0%	EC	EC
		Adept in reading and understanding multi-discipline construction plans and specifications	2	5	4.47	5.00	5.00	0.70	0.83	4	5	1	60.00%	33.33%	0.00%	6.67%	0.00%	93%	7%	EC	EC
		Ability with multiple BIM applications	3	5	4.40	4.00	4.00	0.40	0.63	4	5	1	46.67%	46.67%	6.67%	0.00%	0.00%	93%	0%	ESA	ESA
		Have knowledge of parametric object-based design	3	5	4.20	4.00	4.00	0.46	0.68	4	5	1	33.33%	53.33%	13.33%	0.00%	0.00%	87%	0%	TBD	SA
		Perform basic 3D modeling	3	5	4.40	5.00	5.00	0.54	0.74	4	5	1	53.33%	33.33%	13.33%	0.00%	0.00%	87%	0%	ESA	ESA
		Perform advanced modeling, family editing, custom patterns and profiles, and conceptual massing	2	5	4.13	4.00	4.00	0.70	0.83	4	5	1	33.33%	53.33%	6.67%	6.67%	0.00%	87%	7%	TBD	SA
		Apply BIM-based technology to manage the construction process	4	5	4.53	5.00	5.00	0.27	0.52	4	5	1	53.33%	46.67%	0.00%	0.00%	0.00%	100%	0%	ESA	ESA
		Utilize 3D models for reviewing design and constructability, and model coordination	3	5	4.60	5.00	5.00	0.40	0.63	4	5	1	66.67%	26.67%	6.67%	0.00%	0.00%	93%	0%	EC	EC
		Perform BIM-based planning and logistics, scheduling, cost estimating, and safety planning	3	5	4.40	5.00	5.00	0.69	0.83	4	5	1	60.00%	20.00%	20.00%	0.00%	0.00%	80%	0%	EC	EC
	Cloud Computing	Have knowledge of cloud-based BIM platforms	3	5	4.40	5.00	5.00	0.54	0.74	4	5	1	53.33%	33.33%	13.33%	0.00%	0.00%	87%	0%	ESA	ESA
		Use cloud-based systems to access prroject data	3	5	4.53	5.00	5.00	0.41	0.64	4	5	1	60.00%	33.33%	6.67%	0.00%	0.00%	93%	0%	EC	EC
		Employ cloud-based platforms for BIM visualization and for efficient acquisition of applicable information through the web	3	5	4.53	5.00	5.00	0.41	0.64	4	5	1	60.00%	33.33%	6.67%	0.00%	0.00%	93%	0%	EC	EC
		Apply cloud computing for construction collaboration, coordination, and communication	3	5	4.40	4.00	4.00	0.40	0.63	4	5	1	46.67%	46.67%	6.67%	0.00%	0.00%	93%	0%	ESA	ESA
		Store building data during design and construction and over the entire life of the building on the cloud	2	5	4.20	4.00	4.00	0.74	0.86	4	5	1	40.00%	46.67%	6.67%	6.67%	0.00%	87%	7%	ESA	ESA
		Use cloud platforms for Information exchange and mobile access to information	2	5	4.53	5.00	5.00	0.70	0.83	4	5	1	66.67%	26.67%	0.00%	6.67%	0.00%	93%	7%	EC	EC
	Collaborative Environments	Understand technology for collaborative systems	2	5	4.33	4.00	4.00	0.67	0.82	4	5	1	46.67%	46.67%	0.00%	6.67%	0.00%	93%	7%	ESA	ESA
		Ability to work in a computer-supported cooperative environment	2	5	4.20	4.00	4.00	0.74	0.86	4	5	1	40.00%	46.67%	6.67%	6.67%	0.00%	87%	7%	ESA	ESA
		Define clear outline of responsibilities	3	5	4.40	4.00	4.00	0.40	0.63	4	5	1	46.67%	46.67%	6.67%	0.00%	0.00%	93%	0%	TBD	SA
		Develop and adhere to common processes and tools	2	5	4.20	4.00	4.00	0.74	0.86	4	5	1	33.33%	53.33%	6.67%	6.67%	0.00%	87%	7%	ESA	ESA
		Work efficiently within co-located and remote teams and foster virtual teamwork	3	5	4.47	5.00	5.00	0.41	0.64	4	5	1	53.33%	40.00%	6.67%	0.00%	0.00%	93%	0%	ESA	ESA
		Develop trust among team members	3	5	4.20	4.00	4.00	0.60	0.77	4	5	1	40.00%	40.00%	20.00%	0.00%	0.00%	80%	0%	ESA	ESA
		Establish and communicate conflict resolution strategies	3	5	4.27	4.00	4.00	0.50	0.70	3.5	5	1.5	40.00%	46.67%	13.33%	0.00%	0.00%	87%	0%	TBD	PA
		Willingness to share resources among project participants	4	5	4.60	5.00	5.00	0.26	0.51	4	5	1	60.00%	40.00%	0.00%	0.00%	0.00%	100%	0%	TBD	C
	Programming	Have computer science knowledge	2	5	4.20	4.00	5.00	0.89	0.94	4	5	1	46.67%	33.33%	13.33%	6.67%	0.00%	80%	7%	ESA	ESA
		Comprehend core programming languages	3	5	4.20	4.00	4.00	0.60	0.77	4	5	1	40.00%	40.00%	20.00%	0.00%	0.00%	80%	0%	TBD	SA
		Ability to write codes	2	5	3.73	4.00	4.00	0.78	0.88	3.5	4	0.5	13.33%	60.00%	13.33%	13.33%	0.00%	73%	13%	ESA	ESA
		Perform computational design using Dynamo	2	5	3.87	4.00	4.00	0.98	0.99	3	4.5	1.5	26.67%	46.67%	13.33%	13.33%	0.00%	73%	13%	TBD	PA
		Create applications or plug-ins	2	5	3.73	4.00	4.00	0.92	0.96	3	4.5	1.5	20.00%	53.33%	13.33%	13.33%	0.00%	73%	13%	TBD	PA

Fig. 2.9 Round 1 combinative multicriteria evaluation process for computing competency group

the aggregated assessment results in both Round 1 and Round 2. Consensus (C) represents the LOA per definition, where 60% or more of the Delphi panel rated the KSA as very important. SA is defined as the LOA, where the combined adjacent scale of the Delphi panel's responses (either 4 and 5 or 1 and 2) PS was ≥70%. PA represents the LOA, where the combined adjacent scale of the Delphi panel's responses (either 4 and 5 or 1 and 2) PS was ≥60%. SD represents the LOA per definition, where 25% of the Delphi panel rated the KSA as very important, with SD > 1 or IQR > 2. TD represents the LOA per definition, where <60% of the Delphi panel rated the KSA as very important or with SD > 1 or IQR > 2. The results of the evaluation process in Round 2, showing the final-achieved LOAs of all the automation and robotics KSAs, are presented in Figs. 2.9, 2.10, 2.11 and 2.12.

ROUND 2 - DELPHI ASSESSMENT OF CONSTRUCTION ROBOTICS TECHNOLOGIES KSAs																						
KNOWLEDGE, SKILLS AND ABILITIES			MIN	MAX	MEAN	MEDIAN	MODE	VAR	STD	Q1	Q3	IQR	% = 5	% = 4	% = 3	% = 2	% = 1	% 4,5	% 1,2	LOA (R1)	LOA (R2)
COMPUTER SCIENCE COMPETENCY GROUP	Big Data Analytics	Ability to visualize data and communicate	4	5	4.60	5.00	5.00	0.26	0.51	4	5	1	60.00%	40.00%	0.00%	0.00%	0.00%	100%	0%	EC	EC
		Collect, analyze and compare project data for reporting and to determine project feasibility	2	5	4.07	4.00	4.00	0.50	0.70	4	4	0	20.00%	73.33%	0.00%	6.67%	0.00%	93%	7%	ESA	ESA
		Develop project benchmarks that define project success	3	5	4.20	4.00	4.00	0.46	0.68	4	5	1	33.33%	53.33%	13.33%	0.00%	0.00%	87%	0%	TBD	SA
		Collect data for project benchmarks and report as necessary to project stakeholders	3	5	4.47	5.00	5.00	0.55	0.74	4	5	1	60.00%	26.67%	13.33%	0.00%	0.00%	87%	0%	TBD	C
		Identify likelihood of future outcomes based on past project data	3	5	4.00	4.00	4.00	0.57	0.76	3.5	4.5	1	26.67%	46.67%	26.67%	0.00%	0.00%	73%	0%	TBD	SA
	Database Concepts	Have knowledge about computer-based databases	2	5	3.93	4.00	4.00	0.78	0.88	3.5	4.5	1	26.67%	46.67%	20.00%	6.67%	0.00%	73%	7%	ESA	ESA
		Examine large databases to get actionable insights	2	5	3.93	4.00	4.00	0.78	0.88	3.5	4.5	1	26.67%	46.67%	20.00%	6.67%	0.00%	73%	7%	TBD	SA
		Extract concepts, interrelationships, and patterns of interest from large databases	3	5	4.40	5.00	5.00	0.69	0.83	3.5	5	1.5	60.00%	20.00%	20.00%	0.00%	0.00%	80%	0%	TBD	PA
		Analyze large amount of construction project data to identify valid, useful, and previously unknown patterns	3	5	4.13	4.00	4.00	0.55	0.74	3	4.5	1.5	33.33%	46.67%	20.00%	0.00%	0.00%	80%	0%	TBD	PA
	Internet of Things	Identify use cases for IoT in construction	2	5	4.27	5.00	5.00	0.92	0.96	4	5	1	53.33%	26.67%	13.33%	6.67%	0.00%	80%	7%	ESA	ESA
		Export BIM content to various formats for use in IoT applications	2	5	3.80	4.00	4.00	0.89	0.94	3.5	4	0.5	20.00%	53.33%	13.33%	13.33%	0.00%	73%	13%	ESA	ESA
		Install sensors on physical devices, and program sensors and connect to an online system	3	5	4.27	4.00	4.00	0.35	0.59	4	5	1	33.33%	60.00%	6.67%	0.00%	0.00%	93%	0%	TBD	SA
		Monitor sensors and devices to produce and communicate reports based on sensor data and feedback	2	5	4.20	4.00	4.00	0.60	0.77	4	5	1	33.33%	60.00%	0.00%	6.67%	0.00%	93%	7%	TBD	SA
		Ability to store data securely	2	5	3.80	4.00	4.00	0.89	0.94	3.5	4	0.5	20.00%	53.33%	13.33%	13.33%	0.00%	73%	13%	ESA	ESA
	Machine Learning	Understand concepts of machine learning and artificial intelligence	3	5	4.33	4.00	4.00	0.38	0.62	4	5	1	40.00%	53.33%	6.67%	0.00%	0.00%	93%	0%	TBD	SA
		Have knowledge of computer science fundamentals - data structures, algorithms, computability and complexity, computer architecture	3	5	4.20	4.00	4.00	0.46	0.68	3.5	5	1.5	33.33%	53.33%	13.33%	0.00%	0.00%	87%	0%	TBD	PA
		Perform data modeling and evaluation	3	5	4.27	4.00	4.00	0.35	0.59	4	5	1	33.33%	60.00%	6.67%	0.00%	0.00%	93%	0%	TBD	SA
		Apply probability and statistics to learn from data and provide insights to build intelligent applications	2	5	4.13	4.00	4.00	0.84	0.92	3.75	5	1.25	40.00%	40.00%	13.33%	6.67%	0.00%	80%	7%	TBD	PA
		Apply machine learning algorithms and libraries	2	5	3.73	4.00	4.00	0.24	0.49	3.5	5	1.5	33.33%	66.67%	0.00%	0.00%	0.00%	100%	0%	TBD	PA
		Configure robotic process automation to automate workflows and perform tasks	2	5	4.33	4.00	4.00	0.67	0.82	4	5	1	46.67%	46.67%	0.00%	6.67%	0.00%	93%	7%	TBD	SA

Fig. 2.10 Round 1 combinative multicriteria evaluation process of the computer science competency group

ROUND 2 - DELPHI ASSESSMENT OF CONSTRUCTION ROBOTICS TECHNOLOGIES KSAs																						
KNOWLEDGE, SKILLS AND ABILITIES			MIN	MAX	MEAN	MEDIAN	MODE	VAR	STD	Q1	Q3	IQR	% = 5	% = 4	% = 3	% = 2	% = 1	% 4,5	% 1,2	LOA (R1)	LOA (R2)
TECHNOLOGY COMPETENCY GROUP	Reality Computing	Use reality capture tools for construction applications	4	5	4.40	4.00	4.00	0.26	0.51	4	5	1	40.00%	60.00%	0.00%	0.00%	0.00%	100%	0%	ESA	ESA
		Understand scopes of work and how reality capture may benefit the project	3	5	4.40	4.00	4.00	0.40	0.63	4	5	1	46.67%	46.67%	6.67%	0.00%	0.00%	93%	0%	ESA	ESA
		Use reality capture tools to collect as-built conditions	4	5	4.60	5.00	5.00	0.26	0.51	4	5	1	60.00%	40.00%	0.00%	0.00%	0.00%	100%	0%	EC	EC
		Integrate 3D models with reality capture information	3	5	4.60	5.00	5.00	0.40	0.63	4	5	1	66.67%	26.67%	6.67%	0.00%	0.00%	93%	0%	EC	EC
		Ensure point clouds and 3D models are coordinated and accurate	4	5	4.47	4.00	4.00	0.27	0.52	4	5	1	46.67%	53.33%	0.00%	0.00%	0.00%	100%	0%	ESA	ESA
		Use reality capture software to extract datasets from images or laser scans	4	5	4.53	5.00	5.00	0.27	0.52	4	5	1	53.33%	46.67%	0.00%	0.00%	0.00%	100%	0%	ESA	ESA
		Understand drone use, insurance requirements, FAA regulations	3	5	4.40	5.00	5.00	0.54	0.74	4	5	1	53.33%	33.33%	13.33%	0.00%	0.00%	87%	0%	ESA	ESA
	Visualization	Understand visualization technologies and their applications in construction	4	5	4.53	5.00	5.00	0.27	0.52	4	5	1	53.33%	46.67%	0.00%	0.00%	0.00%	100%	0%	ESA	ESA
		Develop immersive/interactive visualization environments and interact in them	3	5	4.40	4.00	4.00	0.40	0.63	4	5	1	46.67%	46.67%	6.67%	0.00%	0.00%	93%	0%	ESA	ESA
		Import 3D models into immersive/interactive environments	3	5	4.40	4.00	4.00	0.40	0.63	4	5	1	46.67%	46.67%	6.67%	0.00%	0.00%	93%	0%	ESA	ESA
		Use visualization technologies for design and constructability review	4	5	4.47	4.00	4.00	0.27	0.52	4	5	1	46.67%	53.33%	0.00%	0.00%	0.00%	100%	0%	ESA	ESA
		Perform renders and visualizations of 3D models	3	5	4.47	5.00	5.00	0.55	0.74	4	5	1	60.00%	26.67%	13.33%	0.00%	0.00%	87%	0%	EC	EC
	Automation	Have knowledge of robotics and automation trends and applications in construction	4	5	4.60	5.00	5.00	0.26	0.51	4	5	1	60.00%	40.00%	0.00%	0.00%	0.00%	100%	0%	EC	EC
		Identify use cases for using robotics in construction	4	5	4.67	5.00	5.00	0.24	0.49	4	5	1	66.67%	33.33%	0.00%	0.00%	0.00%	100%	0%	EC	EC
		Understand the basic principle of robotic and automation systems	3	5	4.33	4.00	4.00	0.38	0.62	4	5	1	40.00%	53.33%	6.67%	0.00%	0.00%	93%	0%	ESA	ESA
		Develop 3D models with supplemental detailing to create building production models	3	5	4.33	4.00	4.00	0.38	0.62	4	5	1	40.00%	53.33%	6.67%	0.00%	0.00%	93%	0%	ESA	ESA
		Export 3D model content to various formats for use in 3D printing, manufacturing, prefabrication, etc.	2	5	3.93	4.00	4.00	0.78	0.88	3.5	4.5	1	26.67%	46.67%	20.00%	6.67%	0.00%	73%	7%	ESA	ESA
		Monitor and update models and drawings based on feedback from production	3	5	4.20	4.00	4.00	0.46	0.68	4	5	1	33.33%	53.33%	13.33%	0.00%	0.00%	87%	0%	TBD	SA
		Program and manipulate robotic systems for specific construction tasks	2	5	4.13	4.00	4.00	0.84	0.92	4	5	1	40.00%	40.00%	13.33%	6.67%	0.00%	80%	7%	TBD	SA
		Use iterative programs to create generative designs that optimize construction	3	5	4.40	4.00	4.00	0.40	0.63	4	5	1	46.67%	46.67%	6.67%	0.00%	0.00%	93%	0%	TBD	SA
		Use artificial intelligence to optimize construction schedule, cost, material cuts and quantities	3	5	4.47	5.00	5.00	0.41	0.64	4	5	1	53.33%	40.00%	6.67%	0.00%	0.00%	93%	0%	TBD	SA

Fig. 2.11 Round 1 combinative multicriteria evaluation process of the technology competency group

ROUND 2 - DELPHI ASSESSMENT OF CONSTRUCTION ROBOTICS TECHNOLOGIES KSAs																						
KNOWLEDGE, SKILLS AND ABILITIES			MIN	MAX	MEAN	MEDIAN	MODE	VAR	STD	Q1	Q3	IQR	% = 5	% = 4	% = 3	% = 2	% = 1	% 4,5	% 1,2	LOA (R1)	LOA (R2)
ATTITUDES COMPETENCY GROUP	Technology Confidence	Familiar and up-to-date with emerging technologies in construction	3	5	4.47	5.00	5.00	0.41	0.64	4	5	1	53.33%	40.00%	6.67%	0.00%	0.00%	93%	0%	ESA	ESA
		Technology-savvy	3	5	4.60	5.00	5.00	0.40	0.63	4	5	1	66.67%	26.67%	6.67%	0.00%	0.00%	93%	0%	EC	EC
		Enthusiastic about the potentials of new technologies	3	5	4.53	5.00	5.00	0.41	0.64	4	5	1	60.00%	33.33%	6.67%	0.00%	0.00%	93%	0%	EC	EC
		Generate or adapt technology to serve the company's needs	3	5	4.47	5.00	5.00	0.41	0.64	4	5	1	53.33%	40.00%	6.67%	0.00%	0.00%	93%	0%	ESA	ESA
	Active Learning	Motivation to continually learn	4	5	4.67	5.00	5.00	0.24	0.49	4	5	1	66.67%	33.33%	0.00%	0.00%	0.00%	100%	0%	EC	EC
		Learn new software and hardware as necessary	4	5	4.60	5.00	5.00	0.26	0.51	4	5	1	60.00%	40.00%	0.00%	0.00%	0.00%	100%	0%	EC	EC
		Ability to adapt knowledge to continuous updates and changes of technologies	4	5	4.53	5.00	5.00	0.27	0.52	4	5	1	53.33%	46.67%	0.00%	0.00%	0.00%	100%	0%	ESA	ESA
		Participate in industry events to stay up-to-date with emergent technologies	4	5	4.40	4.00	4.00	0.26	0.51	4	5	1	40.00%	60.00%	0.00%	0.00%	0.00%	100%	0%	ESA	ESA
		Ability to stay updated and grasp new concepts quickly	4	5	4.67	5.00	5.00	0.24	0.49	4	5	1	66.67%	33.33%	0.00%	0.00%	0.00%	100%	0%	EC	EC
	Judgement and Decision Making	Define problems, challenges, or opportunities for new technologies	4	5	4.67	5.00	5.00	0.24	0.49	4	5	1	66.67%	33.33%	0.00%	0.00%	0.00%	100%	0%	EC	EC
		Generate an array of possible solutions or applications	4	5	4.40	4.00	4.00	0.26	0.51	4	5	1	40.00%	60.00%	0.00%	0.00%	0.00%	100%	0%	ESA	ESA
		Evaluate the costs and benefits, or pros and cons, associated with each technology	4	5	4.47	4.00	4.00	0.27	0.52	4	5	1	46.67%	53.33%	0.00%	0.00%	0.00%	100%	0%	ESA	ESA
		Select the optimum solution or technology option	4	5	4.47	4.00	4.00	0.27	0.52	4	5	1	46.67%	53.33%	0.00%	0.00%	0.00%	100%	0%	ESA	ESA
		Implement the solution/option chosen	4	5	4.80	5.00	5.00	0.17	0.41	5	5	0	80.00%	20.00%	0.00%	0.00%	0.00%	100%	0%	EC	EC
		Assess the impact of the decision and modify the course of action as needed	2	5	4.47	5.00	5.00	0.70	0.83	4	5	1	62.50%	31.25%	0.00%	6.25%	0.00%	94%	6%	EC	EC
	Complex Problem Solving	Identify complex problems inherent in the construction process and review related data	4	5	4.47	4.00	4.00	0.27	0.52	4	5	1	46.67%	53.33%	0.00%	0.00%	0.00%	100%	0%	ESA	ESA
		Develop and evaluate alternatives and implement solutions	4	5	4.53	5.00	5.00	0.27	0.52	4	5	1	53.33%	46.67%	0.00%	0.00%	0.00%	100%	0%	ESA	ESA
		Select the best solution and technology befitting the problem	4	5	4.60	5.00	5.00	0.26	0.51	4	5	1	60.00%	40.00%	0.00%	0.00%	0.00%	100%	0%	EC	EC
	Innovation and Creativity	Be innately curious and creative	4	5	4.67	5.00	5.00	0.24	0.49	4	5	1	66.67%	33.33%	0.00%	0.00%	0.00%	100%	0%	EC	EC
		Discover novel ways to approach specific problems	4	5	4.60	5.00	5.00	0.26	0.51	4	5	1	60.00%	40.00%	0.00%	0.00%	0.00%	100%	0%	EC	EC
		Ability to generate and try out innovative ideas	4	5	4.73	5.00	5.00	0.21	0.46	4.5	5	0.5	73.33%	26.67%	0.00%	0.00%	0.00%	100%	0%	EC	EC
		Be keen to make technical innovations	3	5	4.53	5.00	5.00	0.41	0.64	4	5	1	60.00%	33.33%	6.67%	0.00%	0.00%	93%	0%	EC	EC
		Develop new ideas for and answers to work-related problems through creative and alternative thinking	4	5	4.53	5.00	5.00	0.27	0.52	4	5	1	53.33%	46.67%	0.00%	0.00%	0.00%	100%	0%	ESA	ESA
	Leadership	Identify the training needs of the project team	4	5	4.60	5.00	5.00	0.26	0.51	4	5	1	60.00%	40.00%	0.00%	0.00%	0.00%	100%	0%	EC	EC
		Participate in or present training	4	5	4.47	4.00	4.00	0.27	0.52	4	5	1	46.67%	53.33%	0.00%	0.00%	0.00%	100%	0%	ESA	ESA
		Develop training programs or classes, and training materials	3	5	4.47	5.00	5.00	0.41	0.64	4	5	1	53.33%	40.00%	6.67%	0.00%	0.00%	93%	0%	ESA	ESA
		Be willing to lead, take charge, and offer opinions and direction regarding technology integration	4	5	4.67	5.00	5.00	0.24	0.49	4	5	1	66.67%	33.33%	0.00%	0.00%	0.00%	100%	0%	EC	EC
		Define new workflows using new technology	4	5	4.40	4.00	4.00	0.26	0.51	4	5	1	40.00%	60.00%	0.00%	0.00%	0.00%	100%	0%	ESA	ESA
		Present new technology workflows internally	4	5	4.40	4.00	4.00	0.26	0.51	4	5	1	40.00%	60.00%	0.00%	0.00%	0.00%	100%	0%	ESA	ESA
		Participate in pilot programs utilizing new technology	2	5	4.40	5.00	5.00	0.69	0.83	4	5	1	53.33%	40.00%	0.00%	6.67%	0.00%	93%	7%	ESA	ESA

Fig. 2.12 Round 1 combinative multicriteria evaluation process of the attitudes competency group

ROUND 2 - DELPHI ASSESSMENT OF CONSTRUCTION ROBOTICS TECHNOLOGIES KSAs								
Competency Group	KSA Line Items	Percentage of EC Achieved	Percentage of ESA Achieved	Percentage of C Achieved	Percentage of SA Achieved	Percentage of PA Achieved	Percentage of SD Achieved	Percentage of TD Achieved
Computing	28	25%	46%	4%	14%	11%	0%	0%
Computer Science	20	5%	25%	5%	40%	25%	0%	0%
Technology	21	24%	57%	0%	19%	0%	0%	0%
Attitudes	30	50%	50%	0%	0%	0%	0%	0%

Fig. 2.13 Round 2 combinative multicriteria evaluation process of the computing competency group

The summary of the results of the consensus evaluation in Round 2 are shown in Fig. 2.13 and are as follows:

1. 28 KSAs in the computing competency group were analyzed. Out of the 28 KSAs, 25% achieved a LOA of EC, 46% achieved a LOA of ESA, 4% achieved a LOA of C, 14% achieved a LOA of SA, and 11% achieved a LOA of PA.
2. 20 KSAs in the computer science competency group were analyzed. Out of the 20 KSAs, 5% achieved a LOA of EC, 25% achieved a LOA of ESA, 5% achieved a LOA of C, 40% achieved a LOA of SA, and 25% achieved a LOA of PA.

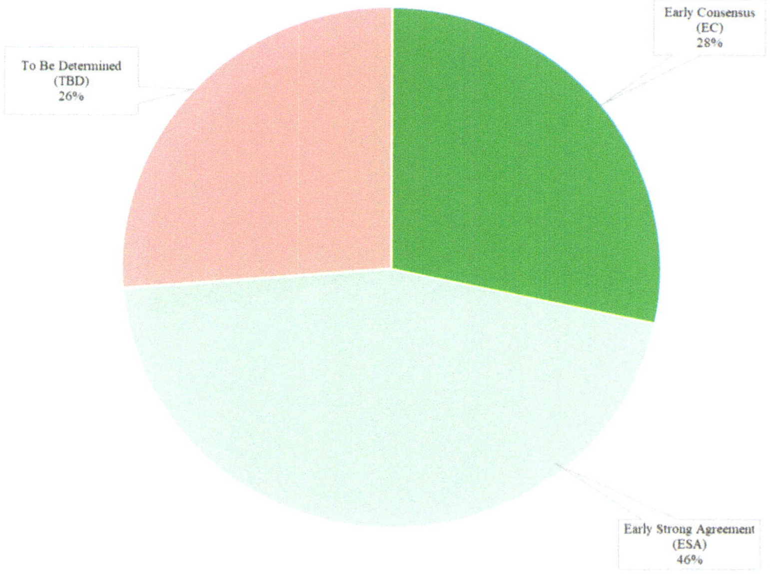

Fig. 2.14 Round 2 combinative multicriteria evaluation process of the computer science competency group

3. 21 KSAs in the technology competency group were analyzed. Out of the 21 KSAs, 24% achieved a LOA of EC, 57% achieved a LOA of ESA, and 19% achieved a LOA of SA.
4. 30 KSAs in the attitudes competency group were analyzed, all of which achieved either EC or ESA LOA in Round 1.

2.4.3.3 Delphi Study Final Results

The Delphi study was conducted to determine a consensus on the foundational KSAs required for automation and robotics proficiency identified during the semi-structured interview phase. A total of 99 KSAs were analyzed, and a consensus was measured with the importance ratings and LOAs attained among the Delphi panelists. Approximately 76% of the KSAs received minimum importance ratings of at least 3 out of 5. The responses of the panelists corroborated with the results from the semi-structured interviews that determined that the KSAs identified were essential for talent cultivation and should be considered for integration in construction education and training.

At the end of the first round of the Delphi study, the panelists reached immediate consensus on the importance of 28 out of the 99 KSAs. Furthermore, the panelists strongly agreed on the importance of 46 out of the 99 KSAs, while the consensus of the importance of the remaining 26 KSAs could not be determined at the end of Round 1. The summary of the results of the percentages of the achieved LOAs after

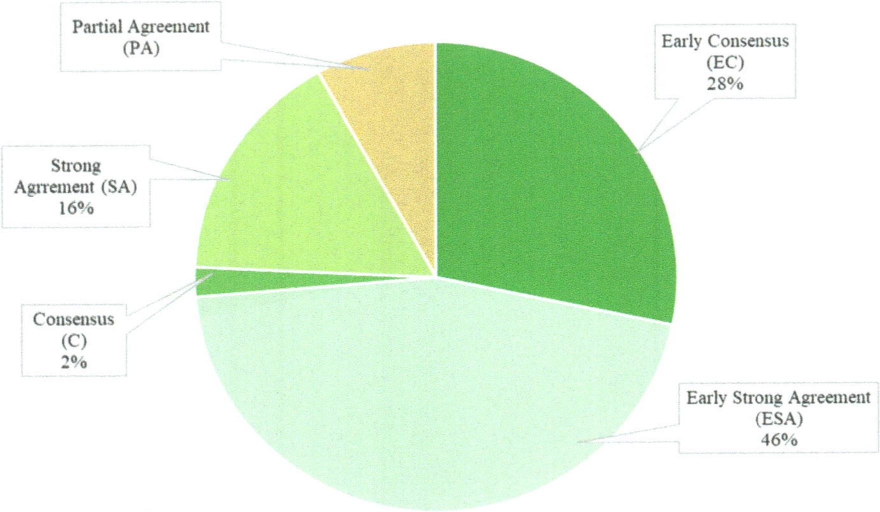

Fig. 2.15 Round 2 combinative multicriteria evaluation process of the technology competency group

the consensus evaluation in Round 1 is shown in Fig. 2.14. The 74 KSAs that attained levels of either EC or ESA were excluded from further evaluation during Round 2 of the Delphi analysis.

The 26 KSAs whose levels of importance were unable to be determined at the end of Round 1 were carried over into Round 2 to be re-rated by the panelists. At the end of the final round of the Delphi study, the panelists reached consensus on the importance of 2 out of the 26 KSAs. Furthermore, the panelists strongly agreed on the importance of 16 out of the 26 KSAs and partially agreed on the importance of the remaining 8 KSAs. The summary of the results of the percentages of the final LOAs after the consensus evaluation in Round 2 is shown in Fig. 2.15.

The results of the Delphi study addressed the research question: What automation and robotics KSAs are essential for construction professionals to perform in the workforce successfully? The results revealed the level to which the panelists agreed on the importance of the automation and robotics KSAs in construction education and training. Although the panelists partially agreed on some of the KSAs, the overall results show that all 99 KSAs need to be addressed and represented in the automation and robotics competency benchmark and performance assessment framework developed in this study.

The most notable limitation encountered during the study was identifying and selecting construction technology and innovation specialists who were the research sample population. Previous research has shown that a minimum of eight panelists is generally recommended for a Delphi study, focusing on expertise, intended outcomes, and an understanding that some participants will not complete the study [14]. To this end, 15 construction technology and innovation specialists who participated in the semi-structured interviews were invited to participate in the Delphi study, and all committed to participate in the study. Furthermore, with the Delphi study conducted via the Internet rather than face to face, there was a lack of direct interaction and discussions with and among the participants, leaving no opportunity for the participants to further elaborate on their views.

2.5 Construction Automation and Robotics Competency Benchmark and Performance Assessment Framework

The automation and robotics competency benchmark and performance assessment framework was developed based on the research findings from the analysis of the automation and robotics KSAs during the semi-structured interview and Delphi study phases. The purpose of the competency benchmark and performance assessment framework is to contribute to automation and robotics curriculum, training, and competency development in industry, as well as academia. The developed benchmark framework is intended to provide measures for construction organizations and academic institutions to evaluate the success of students and new employees in meeting the established set of core automation and robotics competencies during academic and professional training.

2.5.1 Competency Benchmark and Performance Assessment Framework

The competency benchmark and performance assessment framework [21] is designed to be a resource and guideline for training. The users can modify the proposed framework according to organization and training needs. Any additional automation and robotics competency group or KSAs that relate to the specific goals and outcomes of the training program can be added at the discretion of the users. Furthermore, the users of the framework can set their own minimal expected level of competence for each competency group and individual KSAs and overall performance. The benchmarks' ratings aim to determine the level of observed competence needed to meet the standard of the training program.

The automation and robotics competency benchmark framework was organized into the 17 KSA sub-groups and four overarching competency groups determined earlier in the study. The automation and robotics KSAs were further categorized under three training levels that separate the KSAs into a hierarchy based on experience and performance. The three levels of training in the framework are experiential learning, internship, and entry-level. In the context of this study, experiential learning is the training level in the academic setting where students develop automation and robotics KSAs through hands-on experiences combined with learning from

COMPUTING COMPETENCIES

I. BUILDING INFORMATION MODELING (BIM)

Combine graphics expertise with broad construction knowledge to revolutionize the architecture, engineering, construction, and operations (AECO) industry through transformative and collaborative approaches.		
EXPERIENTIAL LEARNING	**INTERNSHIP**	**ENTRY-LEVEL**
Understands BIM concepts and how they apply to construction 1 2 3 4 5 [NO]	Ability with multiple BIM applications 1 2 3 4 5 [NO]	Apply BIM-based technology to manage the construction process 1 2 3 4 5 [NO]
Adept in reading and understanding multi-discipline construction plans and specifications 1 2 3 4 5 [NO]	Utilize 3D models for reviewing design and constructability, and model coordination 1 2 3 4 5 [NO]	Perform BIM-based planning and logistics, scheduling, cost estimating, and safety planning 1 2 3 4 5 [NO]
Perform basic 3D modeling 1 2 3 4 5 [NO]	Have knowledge of parametric object-based design 1 2 3 4 5 [NO]	Perform advanced modeling, family editing, custom patterns and profiles, and conceptual massing 1 2 3 4 5 [NO]

Fig. 2.16 Round 2 combinative multicriteria evaluation process of the attitudes competency group

real-world experiences in the classroom. The internship training level is a structured work experience in organizations that aims to enhance the students' skills through on-the-job training. Lastly, entry-level training is conducted primarily in the industry for new employees to build upon their educational training and previous industry experience. Figure 2.16 is an excerpt of the competency benchmark and performance assessment framework showing the three levels of training.

The performance assessment aspect of the framework entails rating the performance of the trainee in correspondence to each of the automation and robotics KSAs. The performance of the trainees is rated according to a 5-point scale from 1 to 5, with "1" being cognizant, "2" is developing, "3" being competent, "4" being highly skilled, and "5" being expert. The cognizant rating is the proficiency level where the trainee can explain and discuss key concepts of automation and robotics technologies and is familiar with the tasks and processes surrounding their application. The developing rating is the proficiency level where the trainee can integrate automation and robotics technologies with a limited degree of consistency in construction tasks. The competent rating is the proficiency level, where the trainee can integrate automation and robotics technologies with consistency in routine construction tasks. The highly skilled rating is the proficiency level, where the trainee integrates automation and robotics technologies with a high degree of consistency and effectiveness in construction projects. Lastly, the expert rating is the proficiency level where the trainee is exceptionally skillful in demonstrating and integrating automation and robotics technologies in routine and complex construction job tasks and processes.

Depending on the needs of the training program, industry and academic professionals can decide on which competency groups in the competency benchmark and performance assessment framework need to be focused on in the development of competencies. The training's main objective is to select and concentrate on the essential KSAs within each competency group, which the trainee needs to develop. The framework presented in this study is an initial development that is yet to be reviewed and validated by both industry and academic professionals. An initial review of the framework is the next step for continued research in this study, to gather feedback and recommended modifications to ensure consistency automation and robotics training in construction.

2.6 Conclusions

With the rise in the level of advanced automation and robotics technologies being deployed in the construction industry, successful implementation of these technologies requires the inclusion of a workforce with the competencies required to drive automation and robotics processes. To meet the market demand for these professionals, there is a need to integrate automation and robotics technologies and processes in construction curricula. Hence, it is necessary to establish a consensus on the competencies required from construction technology and innovation specialists to ensure proper planning of the integration of automation and robotics in construction education. This study's objective was to determine what fundamental KSAs should be incorporated in construction education by conducting a Delphi study with contributions from the construction technology and innovation specialists.

Upon completion of the semi-structured interviews and two rounds of the Delphi study, four competency groups with a total of 99 KSAs were identified, and a consensus measurement was accomplished for all the KSAs with the LOAs attained between the Delphi panelists. The SD, IQR, and PS of the responses to the importance ratings of the KSAs were calculated and measured for consensus utilizing the combinative multicriteria assessment metrics. The Delphi results showed that not all of the KSAs analyzed achieved full consensus. Out of the 99 KSAs, 28 achieved a LOA of EC, 46 reached ESA, 2 achieved C, 16 achieved SA, and 8 reached a PA consensus level. By identifying these essential KSAs and achieving a consensus on their importance ratings, the contributions from the construction technology and innovation specialists assisted in the development of the competency benchmark and performance assessment framework, to help educators and practitioners in the construction industry develop talent and advance innovation in automation and robotics.

Following the panelists' recommendations from the Delphi study, a preliminary automation and robotics competency benchmarking and performance assessment framework was developed. The framework comprises the developments in personnel trends in the industry, prompted by the implementation of automation and robotics technologies and the essential KSAs linked to them. The competency benchmark framework was organized into 17 competency sub-groups and 4 overarching competency groups, which were further categorized under three training levels. The three levels of training in the framework are experiential learning, internship, and entry-level. The performance assessment in correspondence to each of the automation and robotics KSAs is associated with a 5-point scale from 1 to 5, with "1" being cognizant, "2" is developing, "3" being competent, "4" being highly skilled, and "5" being expert.

The competency benchmark and performance assessment framework is designed to be a resource and guideline for training and is not a mandated policy. Users can modify the proposed framework according to organizational and training needs. The framework presented in this study is an initial development that is yet to be reviewed and validated by both industry and academic professionals. Nevertheless, the framework is expected to lay the foundation for comprehensive development and evaluation of a curriculum to develop students' understanding of how automation and robotics technologies can be used in construction to enhance the levels of performance and productivity of tasks and processes.

Future research in the continuous improvement of the developed competency framework should be conducted to include the following research priorities:

1. Identification of trends in the development of the construction workforce prompted by the application of automation and robotics technologies.
2. Development of a model of student learning outcomes in construction automation and robotics curricula in education and training, with the incorporation of valuable evaluation methods.
3. Benchmarking the desired student learning outcomes and career-specific competencies in developing future construction robotics talent.
4. Evaluation of the possibilities of integrating construction automation and robotics in construction management education curricula.
5. Guiding the development of a training course to ensure the competency benchmark and performance assessment framework provides meaningful outcomes to educators for developing and assessing the competency of students.
6. Testing and validation of the developed competency benchmark and performance assessment framework.

As the field of automation and robotics in construction continues to evolve, the framework is expected to be improved and revised accordingly.

Appendix

References

1. AGC – Associated General Contractors of America, 2019 worker shortage survey analysis.https://www.agc.org/sites/default/files/Files/Communications/2019%20Worker%20Shortage%20Survey%20Analysis.pdf, 2019 (accessed 05.01.19)
2. S. Kim, S. Chang, D. Castro-Lacouture, Dynamic modeling for analyzing impacts of skilled labor shortage on construction project management. J. Manag. Eng. **26**, 04019035 (2020). https://doi.org/10.1061/(asce)me.1943-5479.0000720
3. D. Olsen, M. Tatum, C. Defnall, *How industrial contractors are handling skilled labor shortages in the United States, in: 48th ASC Annual International Conference* (Associated Schools of Construction, Birmingham, UK, 2012) http://ascpro0.ascweb.org/archives/cd/2012/paper/CPGT204002012.pdf

4. F.M. Bademosi, R. Tayeh, R.R.A. Issa, *Skills Assessment for Robotics in Construction Education, in: Workshop on Intelligent Computing in Engineering* (CEUR Workshop Proceedings, Leuven, Belgium, 2019) http://ceur-ws.org/Vol-2394/paper37.pdf
5. S. Bayne, Teacherbot: Interventions in automated teaching. Teach. High. Educ. **20**, 455–467 (2015). https://doi.org/10.1080/13562517.2015.1020783
6. P. Blikstein, Digital fabrication and 'making' in education: the democratization of invention, in *Fablab of Machines, Makers and Inventors*, ed. by J. Walter-Herrmann, C. Büching, (transcript-Verlag, 2014). https://doi.org/10.14361/transcript.9783839423820.203
7. G. Celani, Digital fabrication laboratories: Pedagogy and impacts on architectural education. Nexus Netw. J. **14**, 469–482 (2012). https://doi.org/10.1007/s00004-012-0120-x
8. S. Chong, G.T. Pan, J. Chin, P.L. Show, T.C.K. Yang, C.M. Huang, Integration of 3D printing and industry 4.0 into engineering teaching. Sustain. **10**, 3960 (2018). https://doi.org/10.3390/su10113960
9. S. Khashe, D.J. Gerber, I.F.C. Smith, Surveying the evolution of computing in architecture, engineering, and construction education since 2012. J. Comput. Civ. Eng. **30**, 04016017 (2016). https://doi.org/10.1061/(asce)cp.1943-5487.0000580
10. J. McLurkin, J. Rykowski, M. John, Q. Kaseman, A.J. Lynch, Using multi-robot systems for engineering education: Teaching and outreach with large numbers of an advanced, low-cost robot. IEEE Trans. Educ. **56**, 24–33 (2013). https://doi.org/10.1109/TE.2012.2222646
11. Y. Wang, Y. Chen, X. Tong, Y. Lee, J. Yang, *Robot as a Service in Information Science & Electronic Engineering Education, in: IEEE 13th International Symposium on Autonomous Decentralized Systems (ISADS)* (IEEE, Bangkok, Thailand, 2017). https://doi.org/10.1109/ISADS.2017.53
12. J. Wermann, N. Kliesing, A.W. Colombo, E.C. Moraes, *Impact of New ICT Trends for the Educational Curriculum in the Area of Industrial Automation and Engineering, in: IECON 2015 - 41st Annual Conference of the IEEE Industrial Electronics Society* (IEEE, Yokohama, Japan, 2015). https://doi.org/10.1109/IECON.2015.7392667
13. J. Byrd, B. Spraggon, M. O'Neill, M. Liddy, Could a robot do your job. https://www.abc.net.au/news/2017-08-08/could-a-robot-do-your-job-artificial-intelligence/8782174, 2017 (accessed 17.11.19)
14. M.R. Hallowell, J.A. Gambatese, Qualitative research: Application of the Delphi method to CEM research. J. Constr. Eng. Manag. **136**, 99–107 (2010). https://doi.org/10.1061/(asce)co.1943-7862.0000137
15. L. Giannarou, E. Zervas, Using Delphi technique to build consensus in practice. Int. J. Bus. Sci. Appl. Manag. **9**, 65–82 (2014) https://www.econstor.eu/handle/10419/190657
16. C.C. Hsu, B.A. Sandford, The Delphi technique: Making sense of consensus. Pract. Assess. Res. Eval. **12**, 1–8 (2007) https://doi.org/10.7275/pdz9-th90
17. C.A. Christie, E. Barela, The Delphi technique as a method for increasing inclusion in the evaluation process. Canad. J. Progr. Eval. **20**, 105–122 (2005) https://evaluationcanada.ca/system/files/cjpe-entries/20-1-105.pdf
18. E.A. Jordan, The Semiconductor Industry and Emerging Technologies: A Study Using a Modified Delphi Method, in: ProQuest LLC, D.B.A. Dissertation, University of Phoenix, 2010 (accessed 10.25.19)
19. M.K. Rayens, E.J. Hahn, Building consensus using the policy Delphi method. Policy Polit. Nurs. Pract. **1**, 308–315 (2000). https://doi.org/10.1177/152715440000100409
20. W. Wu, G. Mayo, T.L. McCuen, R.R.A. Issa, D.K. Smith, Building information modeling body of knowledge. II: consensus building and use cases. J. Constr. Eng. Manage. **144**, 04018066 (2018). https://doi.org/10.1061/(asce)co.1943-7862.0001536
21. F.M. Bademosi, R.R.A. Issa, Construction automation and robotics competency benchmark and performance assessment framework. Mendeley Data **V1** (2021). https://doi.org/10.17632/dn2vbdc7wx.1

Chapter 3
Robotics and Automation in Construction (RAC): Priorities and Barriers Toward Productivity Improvement in Civil Infrastructure Projects

Abolfazl Seyrfar, Hossein Ataei, and Ibrahim Osman

Abstract The construction industry, as a large economic sector in industrialized countries, is falling behind other industries in terms of benefitting from automation and robotics to improve its productivity. Therefore, the need for implementing new technologies to resolve construction productivity issues is paramount. By dividing the main research activities of the Robotics and Automation in Construction (RAC) into two large groups, civil infrastructure and residential/commercial buildings, the purpose of this study is to prioritize the areas in which the RAC can be adopted to improve the productivity of civil infrastructure projects. First, different RAC adoption categories in heavy civil projects and main limiting factors were extracted by conducting a literature review. Thereafter, to prioritize the areas with higher potential and need of shifting to automated practices, a structured questionnaire survey was designed to elicit lived experiences of construction professionals with an average of 19 years of experience in heavy civil construction. The results show that the off-site automated prefabrication systems are perceived as the first effective category of RAC adoption, followed by using drones and autonomous vehicles on-site. In addition, top barriers and limiting factors were identified as reluctance of the construction industry to change, high cost of implementation, and lack of government incentive. Overall, it is expected that the findings of this study can help policymakers and construction companies prioritize the investments that are made in RAC implementation in civil infrastructure construction projects. Furthermore, the survey results can further provide insights into the main barriers and challenges in each RAC application area.

Keywords Productivity · Automated construction · Robotics · Off-site construction · Automated prefabrication

A. Seyrfar (✉) · H. Ataei · I. Osman
University of Illinois at Chicago, Chicago, IL, USA
e-mail: aseyrf2@uic.edu; hataei@uic.edu; iosman4@uic.edu

H. Jebelli et al. (eds.), *Automation and Robotics in the Architecture, Engineering, and Construction Industry*, https://doi.org/10.1007/978-3-030-77163-8_3

3.1 Introduction

The construction industry is one of the most important economic sectors in each country [1], and it makes a significant contribution to gross domestic product (GDP) across the world [2, 3]. Although construction productivity is a vital factor in sustaining economic growth [4], it has barely increased and may have even decreased during the last five decades [5]. Moreover, it has been reported that productivity in the construction industry is lower than that in many other industries [6, 7].

The construction industry is falling behind other industries in terms of benefitting from automation and robotics for improving its productivity [8]. As an example, from 1990 to 2000 in Europe, the car manufacturing industry increased its productivity by leveraging automation and robotics six times more than the construction industry [8]. However, due to the increasing complexity of construction and growing attention to the health and safety of the workers, many studies have highlighted the necessity of utilizing new technologies over the past decade [9, 10]. Therefore, robotic and automation has gained more attention in architecture, engineering, and construction (AEC) industry.

In recent years, researchers have attempted to identify the challenges that result in low productivity within the construction industry. Additionally, the role of Robotics and Automation in Construction (RAC) has been explored in order to examine how it can aid in addressing those challenges. The studies have shown that automated systems and robotics have a high potential to enhance the productivity of the construction industry in many different ways [5, 11]. One of the main challenges in construction productivity is the labor productivity decline that has been reported in different economies [2, 12, 13]. As a labor-intensive industry, the construction sector can benefit from robotics and automated systems for improving productivity while reducing labor costs [5]. Another challenge is the occupational health and safety of construction workers, and the RAC can help by reducing injuries and removing workers from dangerous areas [14].

While a large number of studies have been performed to identify and evaluate different types of robotics and automation systems advantages to the construction projects, on-site adoption of RAC still remains restricted and the field is far from mature [9, 15]. Although previous researches highlighted the potential areas in which RAC can be developed and implemented [5, 9], a better understanding of the priorities and barriers of RAC implementation for improving construction productivity is paramount. Therefore, the contribution of RAC toward productivity improvement in different project areas should be carefully evaluated.

In order to achieve this goal, the current study focuses on civil infrastructure projects to prioritize the areas in which the RAC can be adopted to improve construction productivity. This chapter extracts different RAC adoption categories and the main limiting factors from the literature. Thereafter, a structured questionnaire was designed and distributed to construction professionals to find the areas with higher potential and need of shifting to automated practices. Lastly, the survey results were analyzed and presented to fill the gaps in two main aspects: (1) rate the project areas for RAC implementation with regard to productivity improvement and (2) rate major barriers and challenges for RAC adoption.

3.2 Construction Productivity

Productivity is a key factor in evaluating the success of any industry [16]. Based on the scope and level of measurement, productivity in construction can be defined by various terms. However, the main concept for all of them is the same: "more output achievement per resource." Therefore, improving the productivity in the construction industry can be achieved by minimizing the project's cost, duration, material waste, labor, and equipment without negatively affecting neither the quality of output nor the occupational health and safety.

The construction industry suffers from low productivity [17, 18], and it loses $1.6 trillion per annum to the global economy because of low productivity [19]. Productivity issues in the construction industry can be divided into macro- and micro-level. Macro-level issues are mostly related to the contracting methods, labor legislation, and labor organization [20]. Micro-level issues, on the other hand, are dealing with the management and operation of a project, mainly at the job site [20]. The overall productivity of a construction project is mostly governed by the site productivity in the construction phase [21]. Therefore, by elaborating on micro-level issues, previous studies have highlighted the outdoor, labor-intensive nature of construction projects as the most principal cause of low productivity in construction projects [17, 22, 23].

In order to analyze the productivity, the construction productivity factors can be analyzed within different levels such as industry, project, and activity [24, 25]. In this chapter, we will focus on the applications of automated systems and robotics on improving construction productivity at the project level, which can easily be linked to construction activities.

3.3 Robotics and Automation in Construction (RAC)

Construction automation can be defined as "a machine-centered construction factory technology for applying robotic systems on the construction field" [26, 27]. The application of robots in the construction industry has been explored since the early 1980s [28, 29]. Since then, studies have shown that robotics and automated systems can provide many advantages to the construction industry [5, 26, 30]. The main research activities of the Robotics and Automation in Construction (RAC) can be divided into two main groups: civil infrastructure (e.g., road pavers' sensor-based guidance, infrastructure inspection, etc.) and residential/commercial buildings (e.g., building skeleton erection and assembly, concrete compaction, etc.). In this study, the focus is on the applications of RAC in heavy civil infrastructure projects.

In terms of the civil infrastructure projects, four major application categories are proposed: (1) off-site automated systems, (2) on-site automated and robotics systems, (3) drones and autonomous vehicles, and (4) exoskeletons [5]. Off-site systems are mostly related to the prefabrication of concrete and steel elements outside the construction field. On-site systems, on the other hand, cover a broad range of applications that can be implemented directly in the construction field, e.g., inspection, concrete distribution, pipe welding, painting, cleaning, etc. [31]. Compared to off-site RAC adoption, using robotic systems at job sites is more challenging in terms of safety concerns related to human-robot interaction [32, 33]. Therefore, many studies tried to improve the health and safety of construction workers by adopting innovative methods and devices [34, 35]. In the following sections, we will evaluate several applications of each category for productivity improvement in construction projects.

3.4 Research Methodology

In order to assess different application areas of RAC adoption and to highlight the main challenges, a questionnaire survey has been developed. The details of the survey design and target population characteristics are described in the following sections.

3.4.1 Survey Design and Analysis

The questionnaire survey technique is a systematic method of collecting data based on a sample, which has been widely used in construction management research to collect professional views [36–38]. Therefore, a questionnaire survey was administered to extract the experiences and expectations of professionals regarding the applications and challenges of using automation and robotics in civil infrastructure projects in order to improve construction productivity. The population included construction professionals with 19 years of experience (on average) in heavy civil construction. The characteristics of the survey respondents are elaborated in Sect. 3.4.2.

Since no sampling frame was used in this survey, the sample was a non-probability sample [36]. The non-probability sampling has been endorsed as an applicable sampling method when the respondents were not randomly selected from the population but were rather chosen based on their willingness to participate in the study [36, 39, 40].

The main barriers and challenges affecting the implementation of RAC used in this survey (summarized in Table 3.1) were derived from the results obtained by Mahbub [41], Delgado et al. [5], and Buchli et al. [42]. Moreover, the RAC application areas used for the rating in this study were extracted mainly from the results obtained by Delgado et al. [5] including (1) off-site automated prefabrication

Table 3.1 Identified challenges and barriers for the implementation of RAC

Barriers toward implementation of RAC	References
Easy access to labor	[5]
Fragmented nature of the construction industry	[5, 41, 42]
High cost of implementation	[5, 41]
Lack of government incentives	[5]
Lack of strong need to improve productivity	[5]
Reluctances of construction works to change	[5, 41]
Risk of computer problems/on-site accident	[5, 41, 42]
Unproved effectiveness/immature technology	[5, 42]
Untrained workforce	[5, 41, 42]

systems, (2) on-site applications, (3) drones and autonomous vehicles, and (4) exoskeletons.

In order to codify the construction professionals' responses, a 1- to 5-Likert scale was used [43], in which 5 represents the highest importance and 1 represents the lowest importance. Moreover, to verify the clarity of the questions and the structure of the questionnaire, it was pilot-tested by three experts (two from academia and one from the industry).

Finally, to validate the internal consistency of the factors included in the questionnaire, a reliability analysis was performed on the results of the survey. The overall Cronbach's alpha for the RAC application factors and barriers for RAC implementation are 0.90 and 0.92, respectively. Therefore, both Cronbach's alpha values show acceptable internal consistency of the collected data, which makes them suitable for further ranking [44].

3.4.2 *Target Population*

The survey was distributed among 47 professionals in heavy civil construction. Figure 3.1 presents the distribution of respondents' experience in the industry. The target population included a wide range of roles within the construction industry from construction workers to senior construction managers; Fig. 3.2a summarizes the distribution of respondents' roles. Although the majority of the target population were at the bachelor's and master's level, the education level was extended from college degree to the doctorate level (see Fig. 3.2b).

3.5 Results

In this section, the findings of the study are categorized and presented. After analyzing the survey responses, the potential contributions of RAC in productivity improvement, as well as the major barriers and limiting factors, were elaborated.

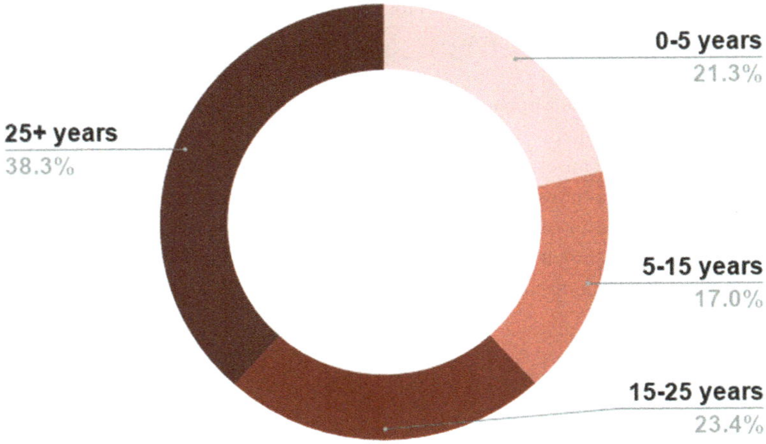

Fig. 3.1 Distribution of the respondents' years of experience

3.5.1 Potential Contribution of RAC in Productivity Improvement

Potential contributions of RAC implementation were evaluated in four different categories: (1) off-site automated prefabrication systems, (2) on-site automated and robotics systems, (3) drones and autonomous vehicles, and (4) exoskeletons [5]. Table 3.2 summarizes the results of the survey questionnaire, where the mean score value of each application was considered as the ranking criteria. Overall, the prefabrication of steel elements gained the highest score in productivity improvement, followed by the prefabrication of concrete elements and using drones and autonomous vehicles to access extreme environments.

3.5.2 Prioritized Project Areas for RAC Implementation

The respondents were asked to rate the efficacy of using robotics and automation in each of the four application categories. Construction professionals rated each category on a scale of 1 to 5, where 1 represents the lowest effectiveness and 5 represents the highest effectiveness. Figure 3.3 summarizes the survey results wherein the majority of the respondents awarded the highest score for using off-site prefabrication systems to improve productivity. In addition, exoskeletons were the last category that the respondents proposed for productivity improvement in construction fields.

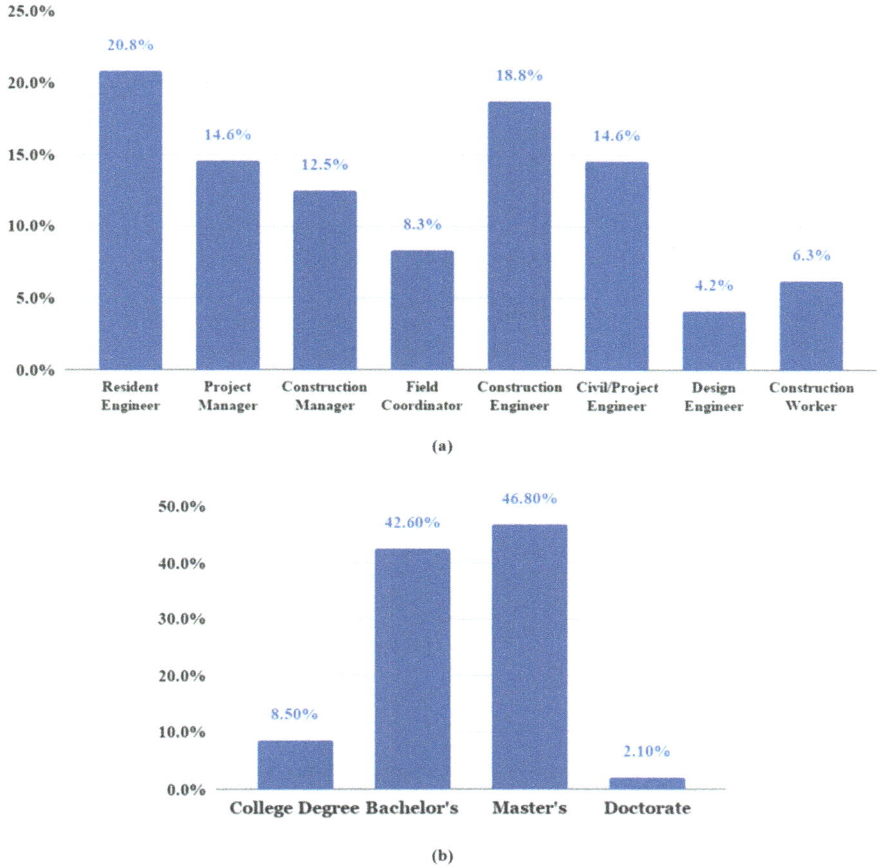

Fig. 3.2 Distribution of target population in (**a**) professional career type and (**b**) level of education

3.5.3 *Main Barriers and Challenges*

Nine potential barriers that might limit the implementation of robotics and automation in civil infrastructure projects were identified through performing a literature review [5, 41, 42]. The respondents were asked to rank these barriers, and the results are shown in Table 3.3. Overall, the professionals selected the lack of government incentive, high cost of implementation, and the reluctance of construction work to change as the three main barriers toward implementing robotics and automated systems in construction projects.

Table 3.2 Descriptive statistics for the subjects

Category	Application	Rank	Mean score
Off-site automated prefabrication	Prefabrication of steel trusses	1	4.34
	Prefabrication of concrete elements	2	4.27
	Prefabrication of wood structural elements	3	4.12
On-site automated and robotics	Welding the pipes and steel elements	1	4.15
	Concrete pouring, distribution, and leveling	2	4.08
	Earthwork (excavation, road compacting, and grading)	3	3.96
	Cleaning process	4	3.95
Drones and autonomous vehicles	Giving access to dangerous environments	1	4.33
	Surveying, monitoring tasks, and bridge inspection.	2	4.15
	Collecting data for progress monitoring	3	3.55
	Autonomous off-road dump trucks	4	3.39
Exoskeleton	Reduce injuries and maintain a healthier workforce	1	4.02
	Solve the ageing construction workforce challenges	2	3.95
	Enable the workers to lift heavy loads and reduce fatigue	3	3.82

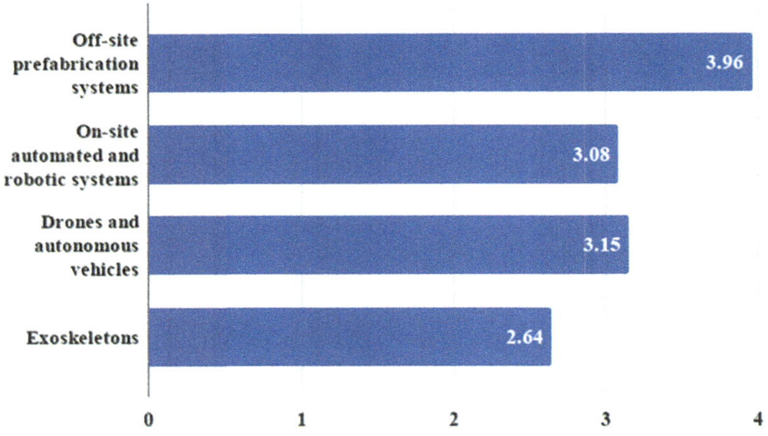

Fig. 3.3 Average score for each application category

Table 3.3 Barriers affecting the implementation of RAC with the rank

Rank	Barrier
1	Lack of government incentives
2	High cost of implementation
3	Reluctances of construction works to change
4	Fragmented nature of the construction industry
5	Unproved effectiveness/immature technology
6	Untrained workforce
7	Risk of computer problems/on-site accident
8	Easy access to labor
9	Lack of strong need to improve productivity

3.6 Discussion

The results show that the majority of the construction professionals surveyed in this study believe that the productivity of civil infrastructure projects should be improved by innovative methods like automation and robotics; hence, the "lack of strong need to improve productivity" was ranked as the weakest barrier. This finding is in agreement with many other studies that proposed low productivity as an issue for heavy civil construction projects [45, 46]. Moreover, the lack of government incentives, high costs of implementation, and reluctance of the construction industry to change were perceived as the strongest barriers toward implementing RAC. Several other studies [5, 41] have also shown the high cost of implementation as the strongest limiting factor for RAC.

The findings of the presented study prioritize off-site automated prefabrication systems rather than other on-site automated systems as the most efficient and effective category of using RAC to improve productivity. The possible reasons for this outcome can be the high risk and cost of on-site implementation and the perception of respondents from the construction field as a reluctant environment to change. Nonetheless, among all the on-site applications, welding the steel elements and using drones to access dangerous areas were identified as the most effective applications of RAC, which indicates the concern of the experienced professionals involved in this study about the safety and risks that exist in construction sites.

All in all, the present study contributes to the body of knowledge by prioritizing different project areas in which RAC can improve construction productivity. In addition, the main challenges of RAC implementation were ranked. The main difference between this study and previous similar studies [5, 41, 42] is that the present study focuses on the RAC contribution to the productivity improvement of civil infrastructure projects.

However, there are some limitations that exist in this study. Although the survey respondents were selected from a wide range of backgrounds, a larger sample size might deliver more reliable results. Moreover, the data collected from construction professionals in one country (USA). However, the characteristics of the construction industry and projects may vary from one country to another. Therefore, further studies should be performed to evaluate the priorities and challenges in other countries.

3.7 Conclusions

Considering productivity as a major factor in evaluating the success of any industry, the need for implementing new technologies and innovations to improve the productivity of construction projects seems paramount. In this chapter, various application areas of implementing robotic and automation in civil infrastructure projects were evaluated by eliciting the real-world experience of construction professionals. The target population was asked to assess the four broad categories of RAC implementation and also rank the major limiting factors in this regard. The insights gained in this study can help the stakeholders of civil infrastructure projects better understand the potential risks and challenges in the implementation of robotics and automated systems. Moreover, the findings can assist project decision-makers in focusing and prioritizing their investments on different application areas of RAC based on their effectiveness on productivity.

References

1. K. Hampson, J.A. Kraatz, A.X. Sanchez, *R&D Investment and Impact in the Global Construction Industry* (Routledge, 2014. ISBN 9781138381360). https://doi.org/10.4324/9781315774916
2. A.A. Javed, W. Pan, L. Chen, W. Zhan, A systemic exploration of drivers for and constraints on construction productivity enhancement. Built Environ. Proj. Asset Manag. **8**, 239–252 (2018). https://doi.org/10.1108/BEPAM-10-2017-0099
3. T.D. Oesterreich, F. Teuteberg, Understanding the implications of digitisation and automation in the context of Industry 4.0: A triangulation approach and elements of a research agenda for the construction industry. Comput. Ind. **83**, 121–139 (2016). https://doi.org/10.1016/j.compind.2016.09.006
4. B. Vogl, M. Abdel-Wahab, Measuring the construction industry's productivity performance: Critique of international productivity comparisons at industry level. J. Constr. Eng. Manag. **141**, 4014085 (2015). https://doi.org/10.1061/(ASCE)CO.1943-7862.0000944
5. J.M.D. Delgado, L. Oyedele, A. Ajayi, L. Akanbi, O. Akinade, M. Bilal, H. Owolabi, Robotics and automated systems in construction: Understanding industry-specific challenges for adoption. J. Build. Eng. **26**, 100868 (2019). https://doi.org/10.1016/j.jobe.2019.100868
6. L. Borg, H.-S. Song, Quality change and implications for productivity development: housing construction in Sweden 1990–2010. J. Constr. Eng. Manag. **141**, 5014014 (2015). https://doi.org/10.1061/(ASCE)CO.1943-7862.0000928
7. W. Zhan, W. Pan, Formulating systemic construction productivity enhancement strategies. J. Constr. Eng. Manag. **146**, 5020008 (2020). https://doi.org/10.1061/(ASCE)CO.1943-7862.0001886

8. B. Carlos, A. Mohamed, Trends in Robotics and Automation in Construction, in: Robot. Autom. Constr., InTech, 2008. DOI: https://doi.org/10.5772/5865. Available from: https://www.intechopen.com/books/robotics_and_automation_in_construction/trends_in_robotics_and_automation_in_construction. Accessed 25 Feb 2021
9. S. Cai, Z. Ma, M.J. Skibniewski, S. Bao, Construction automation and robotics for high-rise buildings over the past decades: A comprehensive review. Adv. Eng. Inform. **42**, 100989 (2019). https://doi.org/10.1016/j.aei.2019.100989
10. T. Bock, The future of construction automation: Technological disruption and the upcoming ubiquity of robotics. Autom. Constr. **59**, 113–121 (2015). https://doi.org/10.1016/j.autcon.2015.07.022
11. P.M. Goodrum, C.T. Haas, C. Caldas, D. Zhai, J. Yeiser, D. Homm, Model to predict the impact of a technology on construction productivity. J. Constr. Eng. Manag. **137**, 678–688 (2011). https://doi.org/10.1061/(ASCE)CO.1943-7862.0000328
12. B. Green, Productivity in Construction: Creating a Framework for the Industry to Thrive, Chart. Inst. Build. Brac. UK. (2016). Available from: https://www.ciob.org/industry/research/Productivity-Construction-Creating-framework-industry-thrive. Accessed 25 Feb 2021
13. P.M. Teicholz, Labor-productivity declines in the construction industry: Causes and remedies (a second look). AECbytes Viewp. **67**, 14 (2013) Available from: http://www.aecbytes.com/viewpoint/2013/issue_67.html. Accessed 25 Feb 2021
14. D. Castro-Lacouture, *Construction Automation, in: Springer Handb* (Springer, Autom, 2009), pp. 1063–1078. https://doi.org/10.1007/978-3-540-78831-7_61
15. M.J. Kim, H.-L. Chi, X. Wang, L. Ding, Automation and robotics in construction and civil engineering. J. Intell. Robot. Syst. **79**, 347 (2015). https://doi.org/10.1007/s10846-015-0252-9
16. H.A. Bashir, K. Alzebdeh, A. Al Riyami, Factor analysis of obstacles restraining productivity improvement programs in manufacturing enterprises in Oman. J. Ind. Eng. (2014). https://doi.org/10.1155/2014/195018
17. J. Jang, S. Ahn, S.H. Cha, K. Cho, C. Koo, T.W. Kim, Toward productivity in future construction: Mapping knowledge and finding insights for achieving successful offsite construction projects. J. Comput. Des. Eng. **8**, 1–14 (2021). https://doi.org/10.1093/jcde/qwaa071
18. Y. Teng, C. Mao, G. Liu, X. Wang, Analysis of stakeholder relationships in the industry chain of industrialized building in China. J. Clean. Prod. **152**, 387–398 (2017). https://doi.org/10.1016/j.jclepro.2017.03.094
19. M.W. Barbosa, A. de la C. Vicente, M.B. Ladeira, M.P.V. de Oliveira, Managing supply chain resources with Big Data Analytics: A systematic review. Int. J. Logist. Res. Appl. **21**, 177–200 (2018). https://doi.org/10.1080/13675567.2017.1369501
20. S.P. Dozzi, S.M. AbouRizk, Productivity in Construction, Institute for Research in Construction, National Research Council Ottawa, 1993. ISBN 0-662-21134-0
21. M.-Y. Cheng, M.-T. Cao, A.Y.J. Mendrofa, Dynamic feature selection for accurately predicting construction productivity using symbiotic organisms search-optimized least square support vector machine. J. Build. Eng. **35**, 101973 (2021)
22. N. Lu, R.W. Liska, Designers' and general contractors' perceptions of offsite construction techniques in the United State construction industry. Int. J. Constr. Educ. Res. **4**, 177–188 (2008). https://doi.org/10.1080/15578770802494565
23. T. Wang, Y. Li, L. Zhang, G. Li, Case study of integrated prefab accommodations system for migrant on-site construction workers in China. J. Prof. Issues Eng. Educ. Pract. **142**, 5016005 (2016). https://doi.org/10.1061/(ASCE)EI.1943-5541.0000288
24. A. Hasan, B. Baroudi, A. Elmualim, R. Rameezdeen, Factors affecting construction productivity: A 30 year systematic review. Eng. Constr. Archit. Manag. **25**, 916–937 (2018). https://doi.org/10.1108/ECAM-02-2017-0035
25. S. Dixit, S.N. Mandal, J.V. Thanikal, K. Saurabh, Evolution of studies in construction productivity: A systematic literature review (2006–2017). Ain Shams Eng. J. **10**, 555–564 (2019). https://doi.org/10.1016/j.asej.2018.10.010

26. Q. Chen, B.G. de Soto, B.T. Adey, Construction automation: Research areas, industry concerns and suggestions for advancement. Autom. Constr. **94**, 22–38 (2018). https://doi.org/10.1016/j.autcon.2018.05.028
27. K. Jung, B. Chu, D. Hong, Robot-based construction automation: An application to steel beam assembly (Part II). Autom. Constr. **32**, 62–79 (2013). https://doi.org/10.1016/j.autcon.2012.12.011
28. C. Haas, M. Skibniewski, E. Budny, Robotics in civil engineering. Comput. Civ. Infrastruct. Eng. **10**, 371–381 (1995). https://doi.org/10.1111/j.1467-8667.1995.tb00298.x
29. B.G. de Soto, I. Agusti-Juan, J. Hunhevicz, S. Joss, K. Graser, G. Habert, B.T. Adey, Productivity of digital fabrication in construction: Cost and time analysis of a robotically built wall. Autom. Constr. **92**, 297–311 (2018). https://doi.org/10.1016/j.autcon.2018.04.004
30. V. Gerling, S. Von Mammen, Robotics for self-organised construction, in: 2016 IEEE 1st Int. Work. Found. Appl. Self* Syst. (FAS* W), 162–167 (2016). https://doi.org/10.1109/FAS-W.2016.45
31. K.S. Saidi, T. Bock, C. Georgoulas, *Robotics in Construction, in: Springer Handb* (Springer, Robot, 2016), pp. 1493–1520. https://doi.org/10.1007/978-3-319-32552-1_57
32. D. Kim, A. Goyal, A. Newell, S. Lee, J. Deng, V.R. Kamat, Semantic relation detection between construction entities to support safe human-robot collaboration in construction, in *Comput. Civ. Eng. 2019 Data, Sensing, Anal*, (American Society of Civil Engineers, Reston, VA, 2019), pp. 265–272. https://doi.org/10.1061/9780784482438.034
33. S. You, J.-H. Kim, S. Lee, V. Kamat, L.P. Robert Jr., Enhancing perceived safety in human--robot collaborative construction using immersive virtual environments. Autom. Constr. **96**, 161–170 (2018). https://doi.org/10.1016/j.autcon.2018.09.008
34. C.R. Ahn, S. Lee, C. Sun, H. Jebelli, K. Yang, B. Choi, Wearable sensing technology applications in construction safety and health. J. Constr. Eng. Manag. **145**, 3119007 (2019). https://doi.org/10.1061/(ASCE)CO.1943-7862.0001708
35. H. Jebelli, S. Lee, Feasibility of Wearable Electromyography (EMG) to Assess Construction Workers' Muscle Fatigue, in *Adv Informatics Comput. Civ. Constr. Eng*, (Springer, 2019), pp. 181–187. https://doi.org/10.1007/978-3-030-00220-6_22
36. X. Zhao, B.-G. Hwang, S.P. Low, Critical success factors for enterprise risk management in Chinese construction companies. Constr. Manag. Econ. **31**, 1199–1214 (2013). https://doi.org/10.1080/01446193.2013.867521
37. B.-G. Hwang, E.-S.J. Lim, Critical success factors for key project players and objectives: Case study of Singapore. J. Constr. Eng. Manag. **139**, 204–215 (2013). https://doi.org/10.1061/(ASCE)CO.1943-7862.0000597
38. L.C.M. Tang, B. Atkinson, R.R. Zou, An entropy-based SWOT evaluation process of critical success factors for international market entry: A case study of a medium-sized consulting company. Constr. Manag. Econ. **30**, 821–834 (2012). https://doi.org/10.1080/01446193.2012.697180
39. J.R. Wilkins, Construction workers' perceptions of health and safety training programmes. Constr. Manag. Econ. **29**, 1017–1026 (2011). https://doi.org/10.1080/01446193.2011.633538
40. M.Q. Patton, Two decades of developments in qualitative inquiry: A personal, experiential perspective. Qual. Soc. Work. **1**, 261–283 (2002). https://doi.org/10.1177/1473325002001003636
41. R. Mahbub, *An Investigation into the Barriers to the Implementation of Automation and Robotics Technologies in the Construction Industry* (Queensland University of Technology, 2008) Available from: https://eprints.qut.edu.au/26377/. Accessed 25 Feb 2021
42. J. Buchli, M. Giftthaler, N. Kumar, M. Lussi, T. Sandy, K. Dörfler, N. Hack, Digital in situ fabrication-challenges and opportunities for robotic in situ fabrication in architecture, construction, and beyond. Cem. Concr. Res. **112**, 66–75 (2018). https://doi.org/10.1016/j.cemconres.2018.05.013
43. R. Likert, A technique for the measurement of attitudes. Arch. Psychol. **22**, 1–55 (1932)
44. J.C. Nunnally, Psychometric theory 3E, Tata McGraw-hill education, 1994. ISBN 0071070885, 9780071070881

45. A. Enshassi, S. Mohamed, Z.A. Mustafa, P.E. Mayer, Factors affecting labour productivity in building projects in the Gaza strip. J. Civ. Eng. Manag. **13**, 245–254 (2007). https://doi.org/10.1080/13923730.2007.9636444
46. C.S.A. Siriwardana, J.Y. Ruwanpura, A conceptual model to develop a worker performance measurement tool to improve construction productivity, in *Constr. Res. Congr. 2012 Constr. Challenges a Flat World*, (2012), pp. 179–188. https://doi.org/10.1061/9780784412329.019

Chapter 4
Construction Automation and Sustainable Development

Mohammadsoroush Tafazzoli

Abstract The rising concerns about the adverse consequential effects of the construction industry on the triple bottom lines of sustainability have urged a paramount demand for automation and technology to supersede conventional methods. Considering the substantial applications of automation in the construction industry, there are conceivable potentials to strategize and formulate the capabilities of automation methods to fortify their contributions to sustainability. Such an approach requires a comprehensive understanding of what automation offers and what the construction industry proposes to become more sustainable. This chapter aims to illuminate the connection between automation and sustainable construction by shedding light on the capabilities of automation. The potential contribution of automation to construction is discussed in the bottom lines of sustainability, namely, environment, society, and economy. The chapter will ultimately examine the barriers to the application of automation in the construction industry. Additionally, it will provide some policy recommendations based on existing research on how to progress automation to create safe, productive, environmental-friendly, as well as human-friendly construction approaches.

Keywords Automation · Sustainability · Productivity · Safety · Controllability

4.1 Introduction

One of the major incentives to make fundamental changes in conventional construction methods is comparing them with manufacturing industries. The examination of the two industries reveals that, compared with the manufacturing

M. Tafazzoli (✉)
Washington State University, Pullman, WA, USA
e-mail: tafazzoli@wsu.edu

H. Jebelli et al. (eds.), *Automation and Robotics in the Architecture, Engineering, and Construction Industry*, https://doi.org/10.1007/978-3-030-77163-8_4

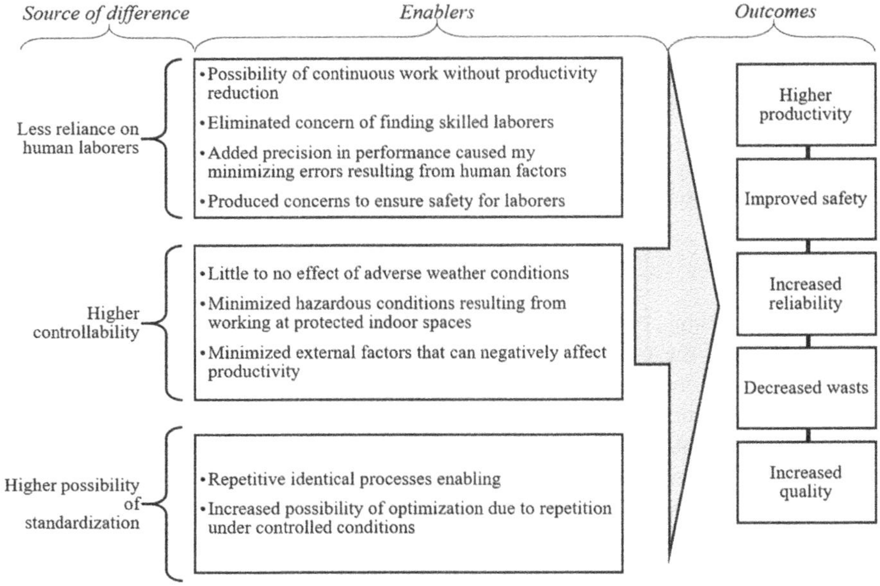

Fig. 4.1 Differences between construction and manufacturing industries' outcomes

industries, the construction industry has significantly less success in achieving safety, productivity, and reliability. The reason for the huge gaps between the performances of the two can be explained in a variety of reasons. Some of them are shown in Fig. 4.1.

The primary reason for the higher success rates of the manufacturing industries is the controllability of the environment and resources resulting from performing the tasks in indoor environments. Additionally, the production occurs in repetitive cycles, by almost unchanging resources, that primarily rely on automatic systems enabling the managers of the system to minimize the risks and reach high levels of productivity and safety.

The massive lost productivity in the construction industry on the one hand and the alarming concerns about lack of materials, high accident rates in construction projects, and substantial negative impacts of this industry on the environment indicate how critical it is to invest in standardizing this industry and improve its controllability [1].

To this end, the fundamental solution is shifting the conventional methods of the construction industry into manufacturing industrialization. Simply put, the more we try to industrialize the construction process, the higher productivity we can achieve. In this regard, Richard [2] divided the degree of industrialization into five levels shown in Fig. 4.2. As it can be seen, automation is a successor of prefabrication and mechanization and is proceeded by robotics and reproduction.

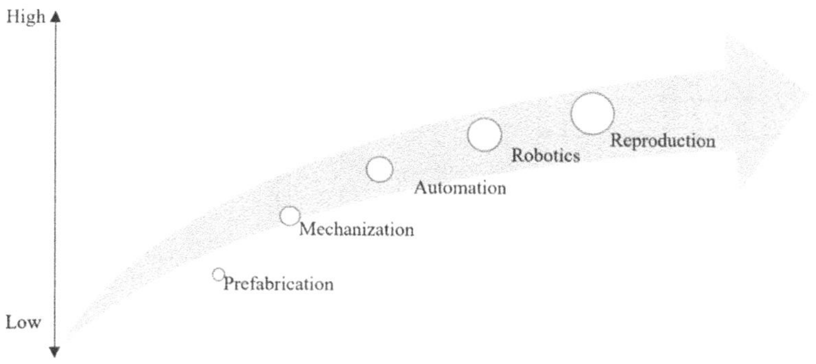

Fig. 4.2 Degree of industrialization

Automation plays an integral role in increasing the controllability and predictability of the construction processes both of which can, directly and indirectly, benefit the triple bottom lines of sustainable development. Shifting the reliance on construction projects from being a labor-intensive industry to automated processes is the next evolution of this industry that has already been launched. Although the pace for the widespread transition from conventional to automated approaches might vary in different countries and regions, the significant rewards of automation are expected to conquer the resistance to change inherited in the construction industry.

Large-scale implementation of automation is expected to create both challenges and opportunities that are unfamiliar to the industry. Some of these aspects have links with sustainable development and its triple bottom lines. The analysis of the connection between automation and sustainability enables us to better prepare for the adoption of the new technologies in construction, mitigate their potential negative impacts, and optimize their utilization. This process is shown in Fig. 4.3.

As illustrated by Fig. 4.3, to optimize the process of automation and align it with sustainability goals, it is essential to profoundly understand and analyze the process by answering the six questions that shed light on possible opportunities and challenges in this transition. This chapter will focus on this goal by discussing the sustainability advantages of automation.

It should be noted that considering that both construction automation and sustainable construction are fairly new transformations that are evolving quickly, the connection between these two topics could potentially have other aspects that are not included in this chapter. The goal of this chapter is to reflect on what is known about the positive impacts of automation on sustainable construction, and some possible negative impacts, such as reducing job opportunities for construction workers, have not been investigated.

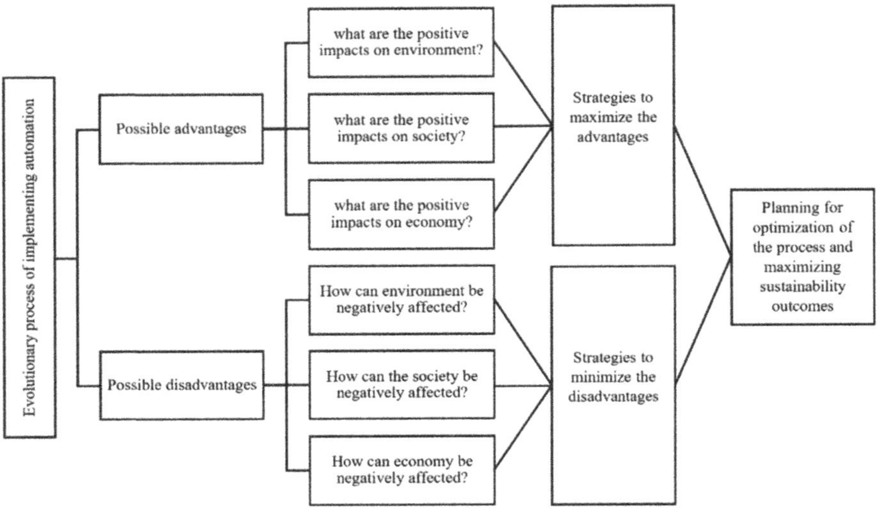

Fig. 4.3 How the evolution of automation should be controlled and optimized

4.2 Contributions of Construction Automation to Sustainability

The construction industry has been widely recognized as one of the major contributors to substantial environmental concerns including depletion of non-renewable sources of energy and materials as well as air, water, and noise pollution [3]. The link between this industry and climate change has also been examined in myriad studies [4–6].

It is widely agreed that without making the construction industry sustainable reaching sustainable development is a far-fetched goal. Although the transition from unsustainable conventional methods to efficient and sustainable practices has been already initiated, it is facing various challenges hindering it to reverse the negative impacts of this industry on the triple pillars of sustainability. Any effort that could contribute to this transition is, therefore, expected to have tremendous impacts in accelerating the movement to build a more sustainable future.

From the sustainability standpoint, the application of automation in the construction industry is aligned with the goals of sustainable development in multiple aspects. Considering the widespread rising attention to sustainability, the positive contributions of automation are expected to incentivize its utilization in the upcoming decades. In the following, some of the major contributions of automation to sustainable development are discussed.

4.2.1 Economic Contributions

Based on construction sustainability studies, economic incentives usually act stronger than environmental or social benefits for investors and stakeholders [7]. This indicates that for wider adoption of automation by the industry we must be aware of the importance of cost-effectiveness. One of the major outcomes of construction automation is the improved productivity of the process. The added precision [8] and controllability of the process in addition to minimizing the possibility of human errors can significantly reduce the waste of time and resources and, therefore, contribute to the productivity and efficiency of the processes. The main economic contributions of construction automation are declining costs and improving efficiency. They can both happen in different domains of a project and its resources including materials and labor. In the following, some of the economic contributions of construction automation are discussed in more detail.

4.2.1.1 Decreased Material Costs

According to the research, by generating 35% of the world's solid waste, the construction industry is one of the primary sources of waste worldwide [9]. Construction wastes are categorized into six groups shown in Fig. 4.4.

It can be interpreted from the causes of waster creation that some of these wastes are produced as a result of low precision in handling materials [11] which is inherited in human activities and performance. For instance, wrong cutting policies and overmixing materials can be minimized by increasing the precision in cuttings and measurements.

Considering the immense negative impacts of construction wastes on the economy and environment, many studies have focused on solutions to mitigate waste generation in this industry [12–14]. Off-site fabrication can be effective in reducing construction wastes [15]. Also, using modular design can lead to a drop in material wastes by enabling off-site processing of materials [16].

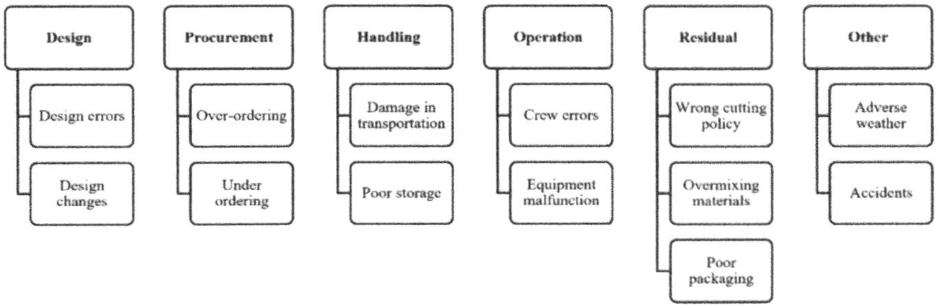

Fig. 4.4 Categories of construction wastes [10]

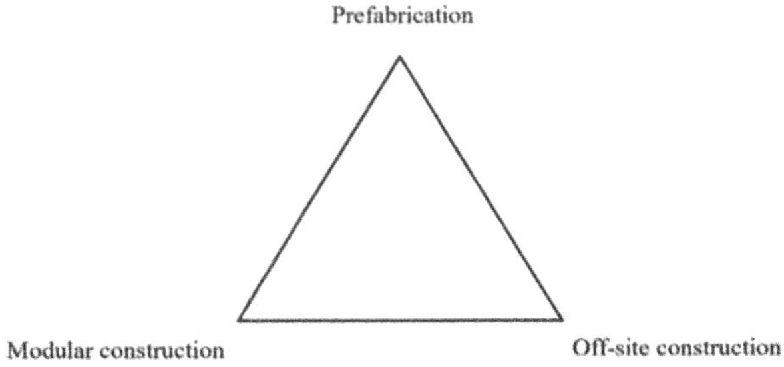

Fig. 4.5 The three major enablers of construction automation

Figure 4.5 shows the enablers of construction automation. All three enablers are associated with reducing material wastes. In a recent study in China, it was shown that prefabrication resulted in 36% and 26% waste reduction for timber formwork and masonry work, respectively [17]. Also, a combination of the modular and coordinated approach of construction can lead to a significant reduction in construction wastes [18].

Despite the common belief, prefabrication and offset construction are not the same. Off-site construction is more focused on performing the tasks at a different location from where the project is being processed, while prefabrication goes beyond location and includes how construction tasks are handled to reduce processing time and effort at the project site.

It should be noted that these approaches could be combined to maximize their positive impacts on sustainability. For instance, modular construction creates an opportunity to utilize the same materials, sizes, and fabrication techniques in large numbers which, in turn, enables us to standardize and industrialize the process. Subsequently, materials can be processed off-site and fabricated before being shipped to the project location. In this case, a combination of all three enablers of construction automation is used.

4.2.1.2 Decreased Labor Costs

Replacing human laborers with robots and other emerging technologies can lead to a significant reduction of the costs that are associated with labor including wages. Figure 4.6 shows the direct and indirect labor costs that can be reduced or eliminated by replacing human laborers with automated technologies. Labor costs are estimated to include between 20% and 50% of a construction project cost. According to [19] indirect labor costs can account for 33–55% of labor costs. Considering the labor intensiveness of this industry, reducing labor costs is expected to substantially reduce project expenses. This can explain why despite the common belief that automation always increases project's costs, it has the potential to reduce these costs depending on the particular costs for the technologies used.

In addition to direct and indirect costs, partial replacement of human laborers by automation and technology can also reduce a variety of issues in the construction

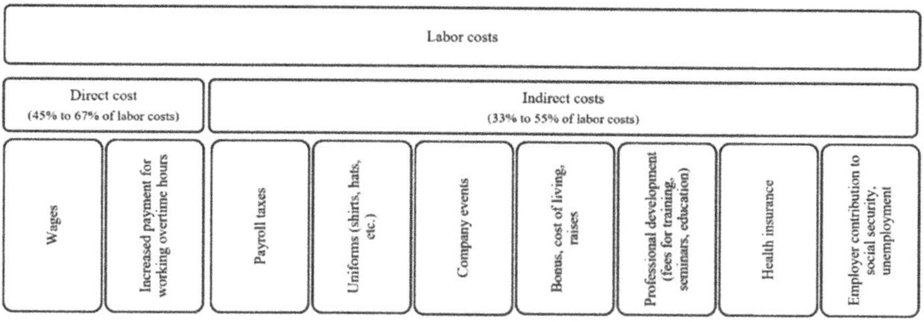

Fig. 4.6 Direct and indirect labor costs reduced or eliminated through automation

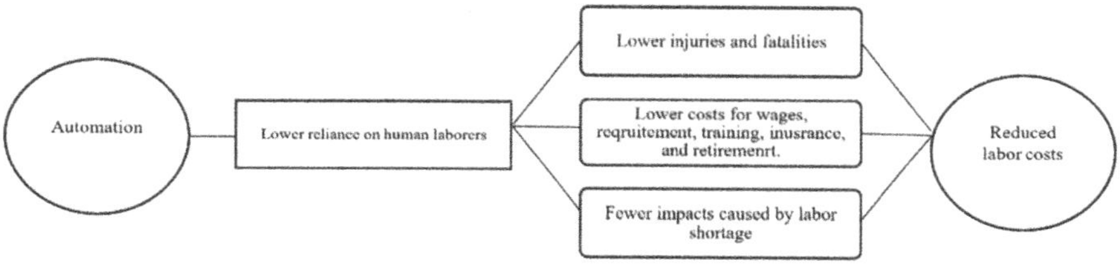

Fig. 4.7 How automation reduces labor costs

industry that have roots in human factors. For example, laborers' age, skill, experience, and motivation are some of the factors affecting laborers' performance. This creates substantial unreliability in the final product based on how the combination of the laborers' strengths and weaknesses in a project can improve or reduce the expected quality. Laborers with insufficient skills can cause considerable costs for rework and fixing defective workmanship.

The cost reduction of construction automation can be attributed to the possibility of continuous work [20]. Additionally, increasing the safety of workers is one of the reasons for robotic construction systems to be promoted in recent years [21]. For instance, the superior precision of robots compared with humans in steel beam assembly work has been shown in research [22]. This chain of impacts is illustrated in Fig. 4.7.

Another contribution of construction automation regarding replacing human laborers is reducing the reliance of the industry on the labor force. The shortage of laborers is one of the major challenges of the construction industry in the United States. This scarcity can sometimes lead to a temporary stop of some construction tasks and delay the project [23, 24].

The shortage of construction workers is more significant in areas with undesirable working conditions such as extremely cold or hot regions, marine projects, and infrastructure projects happening in remote areas. Procuring a workforce for projects in these locations can be extremely more difficult, as the partial higher wages may not incentivize workers to prefer working in difficult conditions.

Contractors may be forced to pay substantially higher wages, and this will have a ripple impact on project costs and consequently the future price of the facility. Implementing automation techniques can potentially reduce the reliance of the project on human laborers and lower the construction costs as well as the cost of the service the facility will provide. These cost reductions can contribute to the economy.

4.2.1.3 Improved Productivity

Increased productivity is a fundamental contributor to economic sustainability. To achieve higher levels of productivity, many aspects of construction including laborer and equipment performance, material handling, building methods, and site management policies should be improved [25].

Decreased reliance on human laborers can increase productivity through (1) improving precision by minimizing human errors; (2) increasing efficiency by minimizing wastes; (3) enabling the company to function in non-working hours, including weekends and night, with minimal cost increase; and (4) uninterrupted and continuous work by eliminating several causes of work stoppage associated with human factors (breaks, using the restroom, etc.).

Maximizing the productivity in construction via automation requires (1) a profound understanding of the capabilities of automation and how a combination of automated technologies can be fitted into the working condition of a construction project and (2) the shortcomings of conventional methods to identify the gaps where automation can serve as a facilitator of the processes. To come up with practical solutions, it is also crucial to consider the cost-effectiveness and possibility of replacing the existing technologies with new ones.

The positive impacts of using new equipment and technology on productivity have been shown in the research [26]. Figure 4.8 shows some of the policies and best practices to improve productivity in construction projects. As seen, productivity is linked with a combination of design, labor, material, tools, and equipment, as well as job site conditions.

Construction automation can potentially have positive impacts on several factors contributing to productivity. Most of these positive impacts are the result of (a) reducing or eliminating human laborers in projects or (b) increasing controllability by standardization of construction tasks.

4.2.1.4 Improved Communications Through Digitization

Construction errors are an indispensable part of almost any project. A lot of research has shown a strong link between construction errors and poor communications [27, 28]. Research indicates a variety of causes for poor communication and also numerous negative impacts of failing to exchange the information on the job site in such a way that the transmitted message means the same to everyone and does not create confusion and misinterpretation or require self-judgment. Three main benefits of

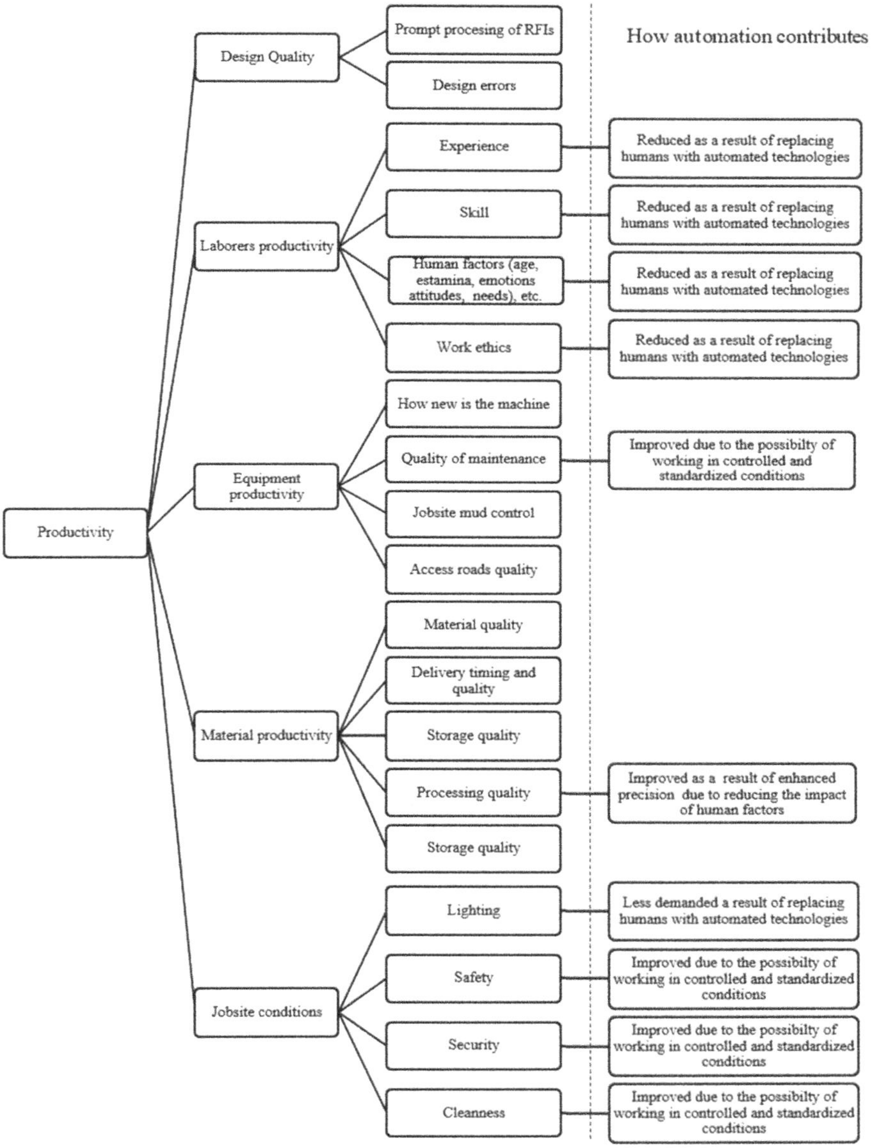

Fig. 4.8 Productivity drivers and how automation can contribute to them

construction automation as contributing to efficiency, improving communication and collaboration, and increasing market share [29]. Replacement of verbal communications with digital commands can minimize the possibility of human errors. Digitalization as a fundamental requirement of automation creates significant opportunities for formulating communications. Automation and digitalization are

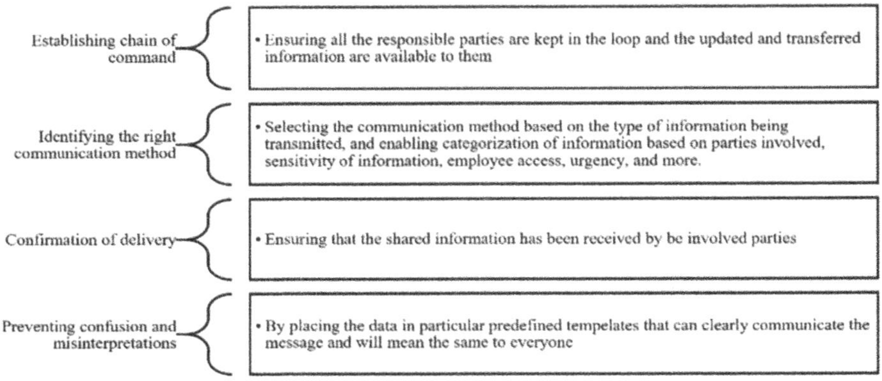

Fig. 4.9 How automation contributes to quality communications

so interwoven that nowadays information modeling and digitalization have proven themselves to be indispensable sections of construction automation [30].

Some automation methods that can contribute to communications in a construction project are shown in Fig. 4.9. Quality communication can potentially prevent the costs resulting from poor communication. These communications can be between the staff or between staff and the robots through codes and the transfer of data that is formulated and is expected to be more precise than the conventional verbal communication between human beings.

4.2.2 *Environmental Contributions*

While a boom in the construction industry is commonly accompanied by multiple social and economic benefits including creating jobs, developing various facilities, enhancing welfare, and more, the environment, as the third bottom line of sustainability, is the one that experiences the most detrimental impacts from the construction development. Automation can mitigate these impacts in a variety of ways. What follows are some of the major contributions of construction automation to the environment.

4.2.2.1 Waste Minimization

As the largest industry in the world, construction projects account for a substantial portion of material waste globally. Bearing in mind that in the upcoming decades shortage of raw materials will become a serious concern of mankind, it is crucial to focus intensively on reducing the waste in the construction process. The review of research [31] indicates several mainstream factors contributing to waste generation shown in Fig. 4.10.

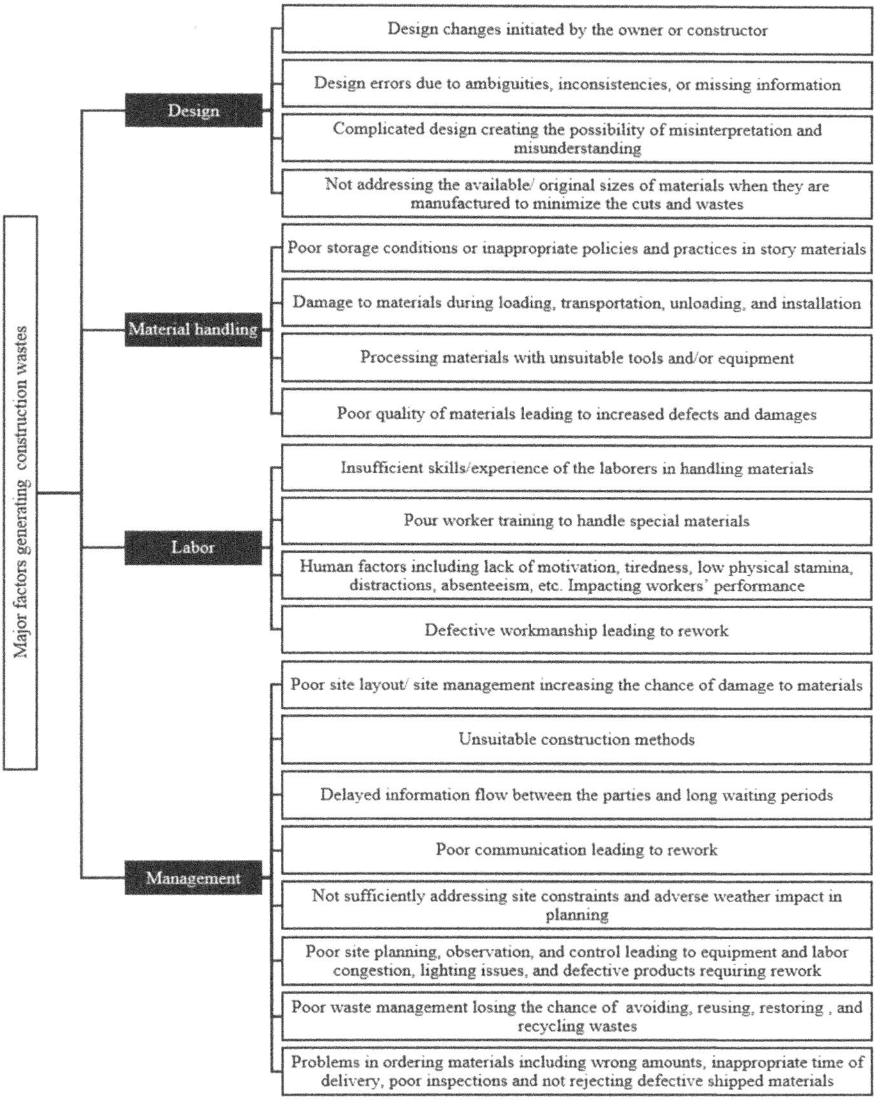

Fig. 4.10 Factors contributing to construction wastes based on categories

Waste minimization through automation happens primarily as a result of increased precision in measurements, manufacturing, assembly, and installation. Each of these processes is prone to be negatively impacted by human errors. Automation minimizes the possibility of such errors by reducing humans' active roles throughout the process. The outcomes of enhanced precision are the efficient use of materials and the reduction of wastes.

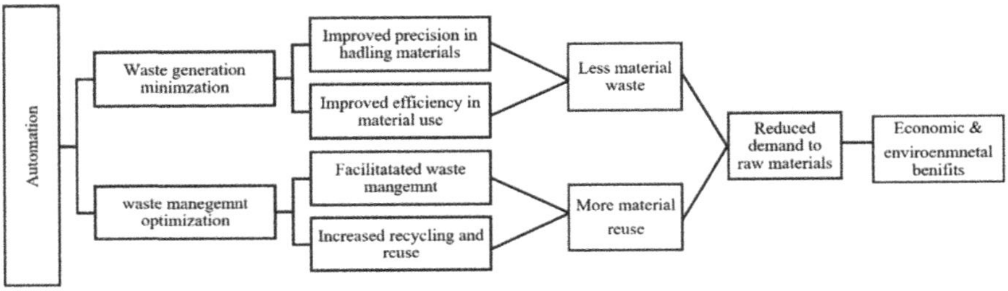

Fig. 4.11 How automation enables the sustainable use of materials

Bearing in mind that the world is expected to face a massive shortage of a variety of non-renewable resources in the upcoming decades, minimizing material waste is expected to become a substantial requirement in the construction industry. Emerging technologies focused on other purposes of construction automation in this domain, such as safety improvement, should be directed to contribute to waste reduction and control for maximizing their subsequent environmental benefits.

In addition to waste reduction, automation can potentially assist in waste management. Reusing and recycling as two of the major approaches in waste control require effective policies and methods in waste management [32, 33]. This includes identification of the source, collection at source, sorting, packaging, and transfer of wasted materials focused on maximizing the volume of materials that can be recycled and reused. Automated construction can potentially enable controllability of waste and therefore the volume of waste that is managed using effective practices (Fig. 4.11).

4.2.2.2 Shrinking Project Duration

Another environmental benefit of construction automation, which has been less discussed in the research, is shortening the project duration. Productivity and duration have an inverse relationship. Therefore, by increasing the productivity of the project, automation can accelerate the construction pace.

The connection between shorter project duration and environmental contribution can be understood by considering the numerous unavoidable disruptions of construction projects to the environment. For example, construction projects commonly generate considerable pollution of air, water, and soil. Although particular preventive measures can mitigate some of these adverse impacts, some damages could be hard or impossible to be eliminated. As another example, during the construction period, the numerous commutes of trucks and heavy equipment to and from the project location can affect the natural soil in the proximity of the project location or transfer chemicals to adjacent streets. The number of such commutes can be reduced when the project is completed more quickly as a result of utilizing automation techniques. Additionally, expediting the project's substantial completion has positive social impacts by reducing the annoyance such as noise and dust pollution for the people and businesses in the vicinity of the project location.

4.2.3 Social Contributions

The social contributions of construction automation have been paid the least attention between the triple bottom lines of sustainability [33]. A closer look at how automation modifies the work environment and the people who are involved in it, however, indicates significant positive social impacts of automation. In the following, some of these major contributions are discussed.

4.2.3.1 Improved Occupational Safety and Health

The construction industry has the highest rank of worksite accidents and fatalities among all the industries. According to the US Center for Construction Research and Training, fatal work injuries in the construction industry account for 17.1% of 4690 injuries that happened in the United States in 2010 accounting for 17.1% of 4690 fatal occupational injuries [34] as cited by [35].

The social impacts of construction accidents are massive and detrimental to this industry. They not only impact the injured individuals and their families but also the contractor and even the community. Accidents can significantly damage the reputation of a company, leading to long-term or even permanent closure of the project, termination of the contract, and economic loss for the stakeholders to name a few. Many of these negative impacts are not numerically quantifiable; however, long-term or permanent injuries will cast a long-lasting shadow on peoples' lives and businesses, and any effort that can potentially mitigate them must be pursued by the industry.

4.2.3.2 Improved Working Environment

Automation can potentially contribute to improving the job environment in multiple aspects. Land use in a project utilizing automation can be optimized by modifying the site layout [36]. Machines and other automation technologies can centralize several activities that would conventionally happen in scattered locations on the job site. By doing so, waste management can also be facilitated which improves the cleanness of the work environment. The connection between worksite cleanness and project outcomes is an observable fact by construction professionals. The positive impact of automation on job sites can therefore improve laborers' motivation and indirectly improve quality and productivity.

Additionally, prefabrication of the materials, which is one of the major tools of automation, can reduce on-site processing of materials. As a result, more space is left to perform construction tasks. The freedom of motion is particularly helpful for construction equipment and in tight working conditions such as building projects in large cities. The elimination of the space required for storing and processing materials, as well as fabricating portions of the structure, can potentially mitigate the risks associated with working in congested conditions and therefore contribute to job site safety.

4.2.3.3 Job Satisfaction

According to the literature, three types of undesirable construction tasks for the laborers are those that are hazardous, repetitive, or happening in unfavorable or untidy settings or to put it simply dull, dirty, and dangerous tasks [22]. By replacing humans with automated technologies, these tasks are reduced which can subsequently lead to job satisfaction.

Moreover, the application of automation can mitigate the risk of exposure to hazardous situations for the laborers. Emerging technologies can improve constructability by mitigating the risk of working under unsafe conditions. A lot of design changes are generated throughout the construction process to improve the constructability of particular portions of the project by making modifications to the original design. Many of these changes require a time-consuming process, may have indirect impacts on project cost or schedule, and can have a negative impact on the client's satisfaction. By lowering the exposure of the laborers to hazardous work conditions, stressful situations can be reduced, and, as a result, job satisfaction of the laborers can be improved. Higher job satisfaction will, in turn, have its own advantages through increasing workers' motivation and the quality of the final product.

4.2.3.4 High-Quality Living Environment

Construction automation can contribute to the quality of the final products in multiple ways (see Fig. 4.12). Automation enables performing tasks that are beyond the capabilities of human laborers. A variety of construction tasks require high precision, physical force, or exposure to hazardous conditions. Automation is a response to compensate for these limitations and produce the final product with the expected quality. Uniformity is the outcome of precession in construction automation and is a key factor in improving the quality of construction. For example, the heat leak in buildings commonly occurs from the spots on the building's shell where proper insulation is disconnected. By enabling uniform construction, automation minimizes fluctuation of quality in different spots of the project.

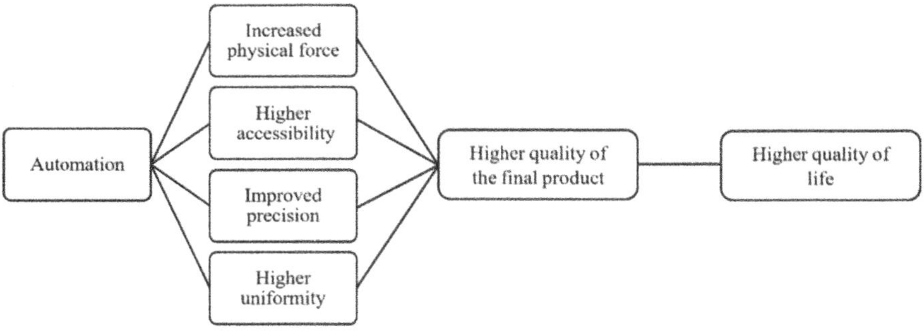

Fig. 4.12 How automation contributes to the quality of life

Low accessibility is a major factor limiting constructability and can happen for a variety of reasons including hazardous situations or narrow access points. The outcome of performing work under difficult circumstances is often compromising quality. Emerging technologies, such as drones that can fly and get access to the points that are hardly reachable for human laborers, enable the constructor to collect and process data which can improve the quality of design and therefore the quality of the final product. A higher quality of buildings is linked to a higher quality of life for its occupants [37].

4.3 Barriers to Maximizing Construction Automation Contribution to Sustainable Development

4.3.1 Lack of Widespread Consensus on Construction Automation Definition

Despite the significant progress of automation, the widespread adoption of construction automation technologies is still an unachieved goal. According to Mahbub [38], the driving force of construction automation in most of the developed countries including North America and Europe are universities, and contractors' willingness to invest in adapting or contributing to the associated technologies is limited.

Considering the gradual formation of automation techniques and robotics in the construction industry, the way they are defined has evolved throughout the years. Automation nowadays encompasses a significantly wider range of construction functions than it would a few decades ago. This has caused a lack of universally agreed definition and therefore scope of construction automation. Simply put, the term construction automation could be used to address a wide variety of technologies with huge differences. In the area of research, studies around construction automation could be sorted into five categories [39] as shown in Fig. 4.13.

Some of the tools that would normally be associated with automation are teleported technologies, utilizing self-controlled, intelligent, or semi-autonomous mechanical or electrical tools or equipment, as well as performing, analyzing, or managing construction activities remotely, collecting an analysis of data to name a few. As can be imagined, each category of these technologies involves different techniques that are connected to dissimilar skills, tools, equipment, requirements, and products. The vast variation which exists between these categories creates a challenge to develop strategies and policies that can effectively contribute to all categories of construction automation. Different associated costs and effectiveness of investment in each category exacerbate the challenge to make comprehensive management policies about automation without breaking them down to category-specific measures. A clear-cut boundary between different available technologies is expected to facilitate addressing them and developing management policies that are particular to each category.

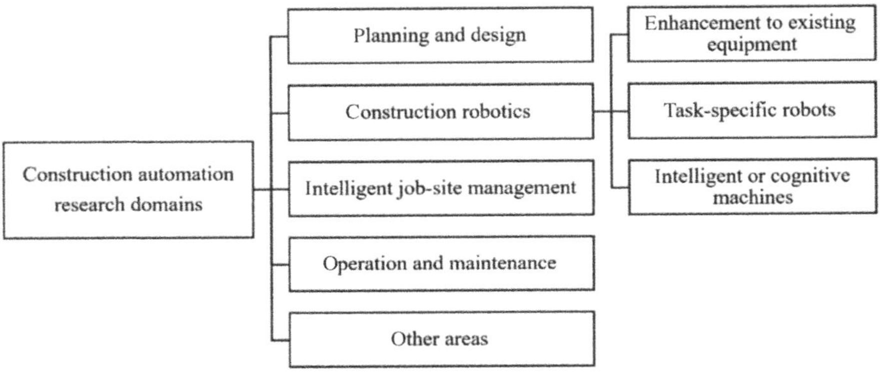

Fig. 4.13 Main categories of construction automation research

4.3.2 Low Demand Due to the High Initial Costs of Automated Production Compared with Human Labor Costs

The high price of automated technologies compared with the traditional workforce is the primary hindrance to their extensive adoption for on-site use [22, 38, 40, 41]. Although human laborers account for 20–50% of project costs, in many cases, they are less expensive than designing, installing, and operating automated processing technologies. While the public's image and assumptions about the associated expenses of automation may be exaggerated or not precise, for many construction tasks that are conventionally done by human laborers, the replacement costs with machines will not make an economic sound. To improve the willingness of the industry in adapting on-site automated construction technologies, we need to invest in reducing the costs and difficulty associated with shipping and installing the associated equipment.

4.3.3 Eliminating Jobs

The common concern about the widespread utilization of automation in the industry is affecting the job market and the possible massive unemployment by replacing human laborers with machines. Some studies have discussed and validated this concern [42–45]. However, a closer look indicates a variety of jobs that are created as a result of automation. Manufacturing, shipping, operating, and maintenance of these machines is a generator of various and numerous new jobs. Nevertheless, the type of these new jobs could be different from the jobs that are eliminated as a result of automation. Thus, creating alternative jobs for the laborers who might lose their jobs as a result of automated processing remains a challenge [46].

It should also be noted that a variety of construction automation technologies function as an addition to existing operations. Some examples are automated

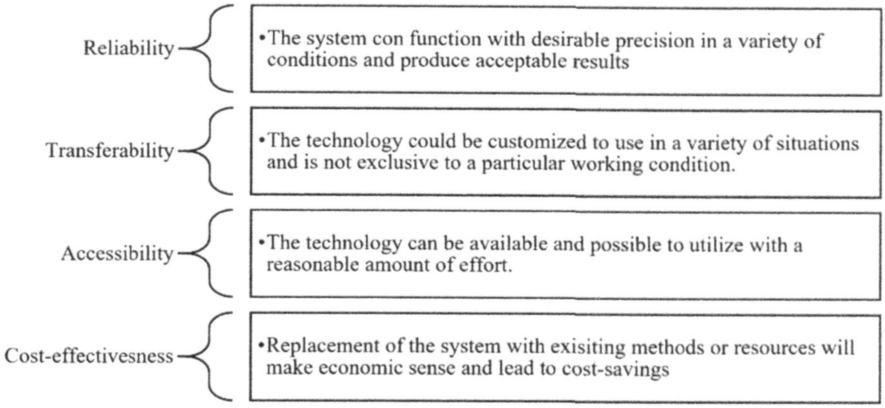

Fig. 4.14 The requirements of a construction automation technology to be widely used

technologies used to improve project administration, monitoring, and control. For instance, an image processing technology that could detect the hazardous situation on a job site is a type of technology that is not intended to physically progress the project and therefore cannot have an impact on the associated jobs.

4.3.4 Lack of Systematic Guidance

The current trend toward the advancement of automation in construction is more focused on information and communication technologies [22]. Significant advancements in image processing such as using drones for monitoring the field have been achieved. However, the high potentials of automation seem to lack effective systematic guidance that can improve its reliability, transferability, accessibility, and cost-effectiveness. The definition of these three recommended approaches is provided in Fig. 4.14. Studying the research and investigating the barriers of automation indicate that wide adoption of automation is far-fetched, particularly for on-site operations, unless these systems evolve and transform in these directions.

4.4 Potentials for Enhancing Construction Automation

Understanding the barriers to construction automation is the prerequisite to lead us to the potential solutions for enhancing its adoption in the industry. These barriers were discussed in the previous part of the chapter. In this part, potentials and opportunities for enhancing construction automation are discussed.

4.4.1 *Prefabrication, Mass Production, and Modular Construction*

One requirement for automation to make economic sense is mass production. Enormous costs of investing in design, manufacturer, installation, training, and maintenance of automated technologies cannot be returned unless the production is pursued in large numbers and long periods. It was mentioned at the beginning of this chapter that industrialization is the key to utilize automation in the construction industry. The uniqueness of project design can hinder creating a template for repetitive processes which is one of the primary elements of industrialization. Modular construction reduces the variation of design as it requires mass production of the units with the same specifications and dimensions. It is therefore an effective response to the process of construction industrialization.

From the lens of sustainable construction, shifting the conventional construction into modular and mass production has numerous benefits and is in perfect alignment with multiple sustainability goals as seen in Fig. 4.15. These benefits indicate why the major goal of making the construction industry more sustainable has close ties with promoting and investing in modular construction. Among these befits, the possibility of reusing materials dismantled from a modular facility is particularly significant as it can contribute to a massive volume of material waste reduction and, therefore, reduced demands for raw materials which is a fundamental environmental concern created by the construction industry.

Prefabrication was conventionally the common definition or understanding of construction automation [22]. Utilizing automation in prefabrication is particularly integral for modular construction that is booming significantly in recent decades. Prefabrication is not limited to modular projects, and this can increase its demand as it can be applied to various types of projects. Figure 4.16 shows the connection between the enablers of construction automation. As seen in this figure, mass housing and modular construction can facilitate using prefabrication methods, and prefabrication is a primary enabler of automated construction.

4.4.2 *Promoting On-Site Use of Automation Technologies*

The uniqueness of design creates a constraint for modular construction which is one of the primary approaches in construction automation. In the meantime, researchers and developers of automated construction technologies believe that the potentials of automation should not be limited to off-site manufacturing and prefabrication.

While prefabrication has been shown to have considerable sustainability advantages, some of the shortcomings inherited in this technology can limit its utilization. One example is the difficulty of making changes in the shipped materials and the time-consuming process of returning the manufactured compartments or replacement of defective pieces. Similarly, adherence to architectural specifications in

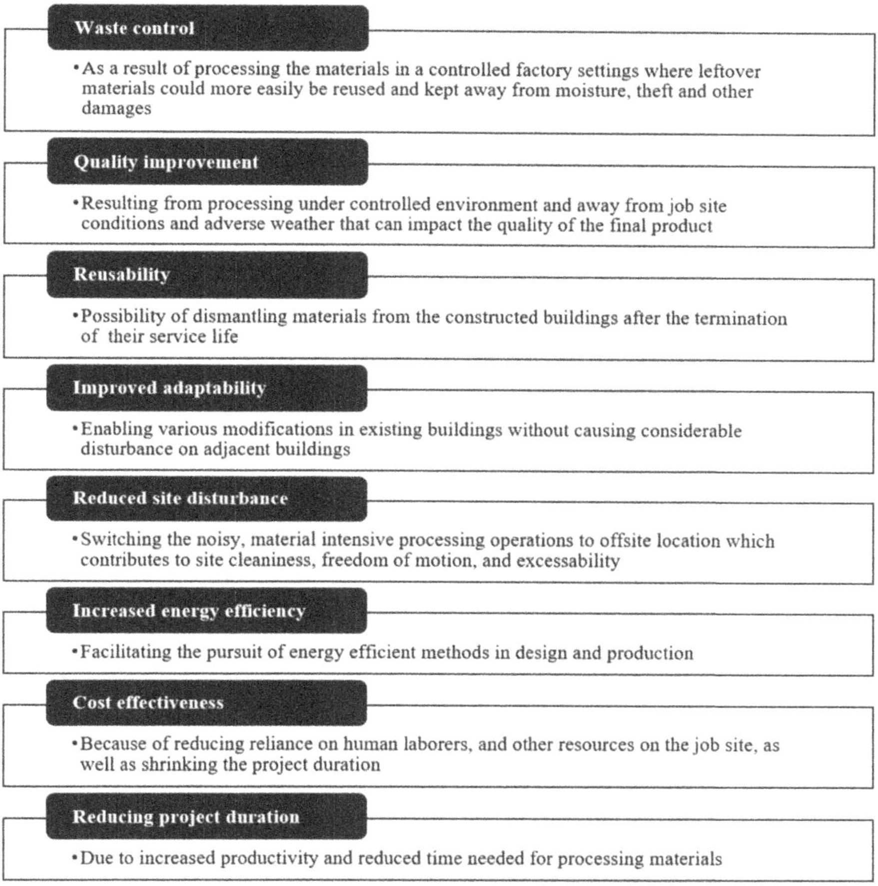

Fig. 4.15 Sustainability benefits of prefabrication and modular construction

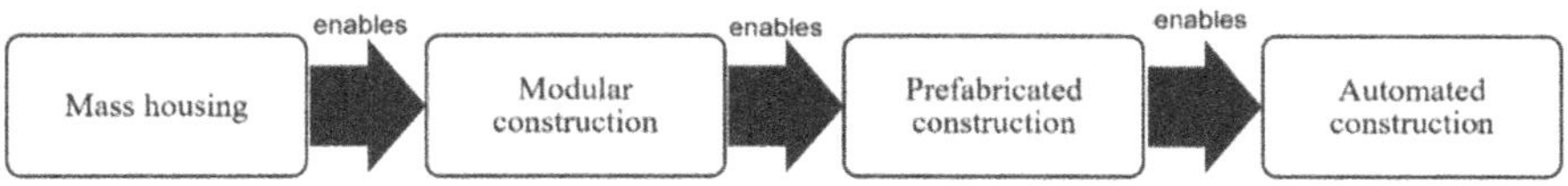

Fig. 4.16 The chain of enablers of automated construction

prefabricated parts and a low level of customizability are some of the other barriers to the widespread adoption of prefabrication.

Limiting automation to off-site production and prefabrication can significantly reduce the positive sustainability impacts of automation, and utilizing automation techniques on-site should remain an indispensable goal in the industry. Despite the fast growth of modular construction, a high percentage of fabrications in construction projects still happen on-site. Moreover, increasing productivity and contributing to safety, as two of the major benefits of construction automation, are more influential in a construction field where there are high potentials for wastes and job site accidents. Applying automation on the project's field can expand the capabilities and positive outcomes of emerging on-site technologies, and at the same time, it will maintain the possibility of customizing compartments on the field where modular construction is not pursued.

To promote on-site utilization of automation in construction, its technology must be directed to improve its flexibility to be used in a variety of operations, as well as accessibility to be transferred to different projects' locations at reasonable expenses. Additionally, the flexibility of these robots to perform a variety of tasks is expected to encourage constructors to be more welcoming to adapting them in their projects.

4.4.3 Improving the Cost-Effectiveness of Automation

In expanding the use of automation in construction, three factors must be taken into consideration: (1) Limiting utilization of construction automation to off-site construction only will not allow us to benefit from many of its advantages on sustainability; (2) for contractors and investors, such technologies are attractive primarily if they can lead to cost-savings; and (3) in the competition to reduce project costs, construction automation has a difficult job to transform and evolve in a way that they can be less expensive than human laborers. The possible conclusion from these facts is that automation must invest in reducing the project costs not only by eliminating human laborers but also by facilitating the entire process of design and construction and performing the operations more cost-effectively. Figure 4.17 shows some of the expected advancements of construction automation in the future.

The limited interest of private owners and investors to contribute to the considerable costs of the transition to automation explains why governmental support is pivotal in accomplishing this goal. As an example, in Europe, the European Union provides funding to projects that utilize automation and prefabrication and therefore have a more efficient energy performance [47].

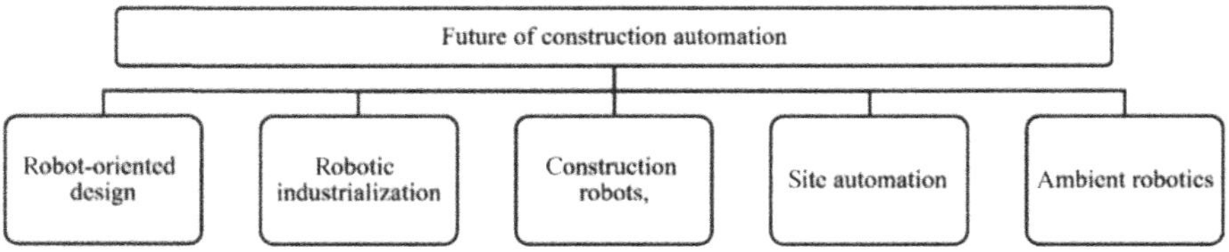

Fig. 4.17 Future of construction automation

4.5 Summary and Conclusion

The poor performance of the construction industry compared with manufacturing industries indicates that increasing controllability and standardization are the keys to making the construction industry more productive, sustainable, and safe. The contributions of construction automation were discussed in each of the three bottom lines of sustainability. The possibility of reducing labor intensiveness through replacing humans with robots was identified as the primary source of these contributions. Through mitigating the possibility of human errors, construction automation can lead to multiple positive impacts on sustainability including added precision to construction activities leading to efficient use of materials and reducing the wastes benefiting both environment and economy; improved safety, healthier, more desirable work environment, and job satisfaction which contribute to social sustainability; and increased productivity that shrinks project duration and benefits all sustainability bottom lines.

The indirect benefits of automation on sustainability were also discussed. It was also argued that the transition to automation in the construction industry which is labor-intensive and resistant to adopt new technologies is a challenge. The barriers of construction automation including higher initial and maintenance costs as well as the lack of skilled individuals who can operate and maintain associated tools and equipment was discussed. It was highlighted that to accelerate the pace of automation in the construction industry, economic incentives act stronger than environmental or social benefits. This means by investing in progressing these technologies to make the construction process more efficient and less costly, automation can be more widely and quickly adopted in the industry. It is important to keep in mind that construction automation is an evolving transition in this industry and by identifying the existing shortcomings of conventional methods and analyzing the potentials of automation; its capabilities can be more effectively directed to resolve the existing issues and make more contributions to the construction industry.

References

1. A. Karji, M. Namian, M. Tafazzoli, Identifying the key barriers to promote sustainable construction in the United States: a principal component analysis. Sustainability **12**(12), 5088 (2020)
2. R.B. Richard, Industrialised building systems: reproduction before automation and robotics. Autom. Constr. **14**(4), 442–451 (2005)
3. M. Tafazzoli, E. Mousavi, S. Kermanshachi, Opportunities and challenges of green-lean: an integrated system for sustainable construction. Sustainability **12**(11), 4460 (2020)
4. A.C. Hurlimann, G. Warren-Myers, G.R. Browne, Is the Australian construction industry prepared for climate change? Build. Environ. **153**, 128–137 (2019)
5. S. Sorrell, Making the link: climate policy and the reform of the UK construction industry. Energy Policy **31**(9), 865–878 (2003)

6. L. Boatman, D. Chaplan, S. Teran, L.S. Welch, Creating a climate for ergonomic changes in the construction industry. Am. J. Ind. Med. **58**(8), 858–869 (2015)
7. M. Tafazzoli, Accelerating the green movement: Major barriers to sustainable construction, in *54th ASC Annual International Conference Proceedings*, (Associated Schools of Construction, 2018)
8. A. Warszawski, *Industrialized and Automated Building Systems: A Managerial Approach* (Routledge, 2003)
9. Hendriks, C. F., & Pietersen, H. S. (2000). Sustainable Raw Materials–Construction and Demolition Waste (165-SRM).. State-of-the-Art Report of RILEM Technical Committee.
10. M. Tafazzoli, A method to measure material-use efficiency in construction projects, in *Proceedings of the 52nd Annual International Conference of Associated School of Construction, Provo, Utah*, (2016)
11. G.A. Bekr, Study of the causes and magnitude of wastage of materials on construction sites in Jordan. J. Construct. Eng. **2014**, 1–6 (2014)
12. Z. Ding, M. Zhu, V.W. Tam, G. Yi, C.N. Tran, A system dynamics-based environmental benefit assessment model of construction waste reduction management at the design and construction stages. J. Clean. Prod. **176**, 676–692 (2018)
13. M.U. Hossain, Z. Wu, C.S. Poon, Comparative environmental evaluation of construction waste management through different waste sorting systems in Hong Kong. Waste Manage. **69**, 325–335 (2017)
14. J. Li, J. Zuo, H. Cai, G. Zillante, Construction waste reduction behavior of contractor employees: An extended theory of planned behavior model approach. J. Clean. Prod. **172**, 1399–1408 (2018)
15. A. Baldwin, C.S. Poon, L.Y. Shen, S. Austin, I. Wong, Designing out waste in high-rise residential buildings: analysis of precasting methods and traditional construction. Renew. Energy **34**(9), 2067–2073 (2009)
16. J. Wang, Z. Li, V.W. Tam, Critical factors in effective construction waste minimization at the design stage: a Shenzhen case study, China. Resour. Conserv. Recycl. **82**, 1–7 (2014)
17. J. Hao, Z. Chen, Z. Zhang, G. Loehlein, Quantifying construction waste reduction through the application of prefabrication: a case study in Anhui, China. Environ. Sci. Pollut. Res. Int. (2020)
18. S. Banihashemi, A. Tabadkani, M.R. Hosseini, Modular coordination-based generative algorithm to optimize construction waste. Procedia Eng. **180**, 631–639 (2017)
19. F.E. Gould, N.E. Joyce, *Construction Project Management* (Prentice Hall, 2009)
20. T. Bock, Construction automation and robotics. Robot. Autom. Constr., 21–42 (2008)
21. B. Chu, K. Jung, M.T. Lim, D. Hong, Robot-based construction automation: an application to steel beam assembly (Part I). Autom. Constr. **32**, 46–61 (2013)
22. M. Pan, T. Linner, W. Pan, H. Cheng, T. Bock, A framework of indicators for assessing construction automation and robotics in the sustainability context. J. Clean. Prod. **182**, 82–95 (2018)
23. S. Kim, S. Chang, D. Castro-Lacouture, Dynamic modeling for analyzing impacts of skilled labor shortage on construction project management. J. Manage. Eng. **36**(1), 04019035 (2020)
24. J.E. Gomar, C.T. Haas, D.P. Morton, Assignment and allocation optimization of partially multiskilled workforce. J. Constr. Eng. Manage. **128**(2), 103–109 (2002)
25. D. Arditi, K. Mochtar, Trends in productivity improvement in the US construction industry. Constr. Manage. Econ. **18**(1), 15–27 (2000)
26. P. Goodrum, C. Haas, Long-term impact of equipment technology on labor productivity in the U.S. construction industry at the activity level. J. Constr. Eng. Manage. **130**(1), 124–133 (2004)
27. Y. Gamil, I.A. Rahman, Identification of causes and effects of poor communication in construction industry: a theoretical review. Emerg. Sci. J. **1**(4), 239–247 (2017)
28. L.G. Tipili, P.O. Ojeba, Evaluating the effects of communication in construction project delivery in Nigeria, in *Proceedings of the Multi-Disciplinary Academic Conference on Sustainable Development*, (2014, July)

29. Q. Chen, B.G. de Soto, B.T. Adey, Construction automation: research areas, industry concerns and suggestions for advancement. Autom. Constr. **94**, 22–38 (2018)
30. T.D. Oesterreich, F. Teuteberg, Understanding the implications of digit-isation and automation in the context of Industry 4.0: a triangulation approach and elements of a research agenda for the construction indus-try. Comput. Ind. **83**, 121–139 (2016)
31. S. Nagapan, I. Abdul Rahman, A. Asmi, *A Review of Construction Waste Cause Factors* (2011)
32. L.Y. Shen, V.W. Tam, C.M. Tam, D. Drew, Mapping approach for examining waste management on construction sites. J. Constr. Eng. Manage. **130**(4), 472–481 (2004)
33. W. Pan, Y. Ning, Dialectics of sustainable building: evidence from empirical studies 1987–2013. Build. Environ. **82**, 666–674 (2014)
34. CPWR (Center for Construction Research and Training), *The Construction Chart Book: The U.S. Construction Industry and its Workers* (CPWR – The Center for Construction Research and Training, Silver Spring, 2013)
35. S.O. Eteifa, I.H. El-Adaway, Using social network analysis to model the interaction between root causes of fatalities in the construction indus-try. J. Manage. Eng. **34**(1), 04017045 (2018)
36. C. Balaguer, M. Abderrahim, J.M. Navarro, S. Boudjabeur, P. Aromaa, K. Kahkonen, et al., FutureHome: an integrated construction automation approach. IEEE Robot. Autom. Mag. **9**(1), 55–66 (2002)
37. S.S. Kamaruddin, M.F. Mohammad, R. Mahbub, Barriers and impact of mechanization and automation in construction to achieve better quality products. Procedia-Social Behav. Sci. **222**, 111–120 (2016)
38. R. Mahbub, An investigation into the barriers to the implementation of automation and robotics technologies in the construction industry. Doctoral dissertation, Queensland University of Technology, 2008
39. Z. Struková, M. Líška, Application of automation and robotics in construction work execution. AD ALTA J. Interdiscip. Res. **2**(2), 121–125 (2012)
40. T. Bock, T. Linner, *Robotic Industrialization: Automation and Robotic Technologies for Customized Component, Module, and Building Prefabrication* (Cambridge University Press, Cambridge, 2015)
41. L. Cousineau, N. Miura, *Construction Robots: The Search for New Building Technology in Japan* (ASCE Publications, 1998)
42. J. Manyika, M. Chui, J. Bughin, R. Dobbs, P. Bisson, A. Marrs, *Disruptive Technologies: Advances that will Transform Life, Business, and the Global Economy*, vol 180 (McKinsey Global Institute, San Francisco, 2013)
43. M. Wolfgang, *The Robotics Market—Figures and Forecasts* (RoboBusiness, Boston Consulting Group, Boston, 2016)
44. E. Brynjolfsson, A. McAfee, *How the Digital Revolution is Accelerating Innovation, Driving Productivity, and Irreversibly Transforming Employment and the Economy* (Digital Frontier Press, 2011)
45. M. Ford, *The Rise of the Robots: Technology and the Threat of Mass Unemployment* (Simon and Schuster, 2015)
46. B. Vermeulen, J. Kesselhut, A. Pyka, P.P. Saviotti, The impact of automation on employment: just the usual structural change? Sustainability **10**(5), 1661 (2018)
47. BERTIM, Building Energy Renovation through Timber Prefabricated Modules (2016), http://www.bertim.eu/

Chapter 5
A Briefing on Technological, Societal, and Economical Aspects of Automation in Construction

Farid Sartipi

Abstract The concept of smart cities and automation in construction are promising ways for managers today to increase the productivity throughout the project development. However, apart from the managerial point of view, there are several other important aspects that require a comprehensive evaluation before a proper implementation of an autonomous system. There must be a technological gap which the automated system is about to tackle, i.e. engineering design usually is accompanied with technological evaluation. The human interaction with robots, also, is another key area which requires precise attention as if a society does not accept the new way of business operation introduced by automation, and the failure of the technological advancement is most likely. Humans and robots are two different creatures with distinctive biological differences, who are supposed to cooperate and maintain/enhance the social harmony. The third important factor is the economic aspects of automation. In that regard, a balance must be made between several economic variables such as employment, cash rate, inflation, etc. In this study, a detailed analysis will be given using case studies (for technological aspects), bibliometric literature review (for social aspects) and econometrics (for economic aspects). To bridge the knowledge gap, this chapter attempts to draw the connections between these three important factors. The results will help the construction managers gain a comprehensive understanding around causality of the areas in which their proposed automation plan is about to be implemented.

Keywords Automation in construction · Image processing · Human-robot interaction

---

F. Sartipi (✉)
Institute of Construction Materials, Kingswood, NSW, Australia
e-mail: farid.sartipi@iconsmat.com.au

H. Jebelli et al. (eds.), *Automation and Robotics in the Architecture, Engineering, and Construction Industry*, https://doi.org/10.1007/978-3-030-77163-8_5

5.1 Introduction

Being technologically passive has been one of the well-known characteristics of the construction industry for many years when it is compared with other industries such as automotive industry, medical sector or even retail. It is almost impossible to find an industry sector without a touch of digital technologies, meanwhile the construction industry is ranked amongst the lowest in the list in terms of technological aversion [1]. And this had come with consequences; high labour intensity, failures caused by the human errors, longer project completion period, health risks, high capital investment and so on [2, 3]. In a world where population is growing at a record high on an exponential scale and the rate of homelessness is skyrocketing, the necessity of rapid housing and infrastructure construction had become more important than ever. And, that is not practicable unless automation and robotics would be embraced by the construction industry.

Although there had been many shortfalls in terms of embracing digital technologies within the construction industry, in the recent years, the concept of *smart cities* in conjunction with the emerging 5G network opens hope for a construction industry with a higher magnitude of automation implemented [4]. A digitally connected built environment, in other words, allows a faster transfer of data and information through the crucial elements of the city. Those elements could be a simple red light at a cross section connected to the network or the speed camera in the highways that are already in existence. A more futuristic outlook, however, can be a road infrastructure capable of constantly charging the electronic vehicles on the go. Such examples are a lot in numbers and will be illustrated throughout the passage.

Focusing on the construction industry and the implication of robotics in construction, as of today, two major technologies are trending: contour crafting (3D printing) and image processing. Other interesting and at the same time important relevant subjects in this cluster are the social interaction of robots with human beings and the economic impacts that they have on the employment rate. Each requires a precise elaboration as these two subjects closely influence the human wellbeing. The rationale behind the structure of this chapter is shown in Fig. 5.1.

In summary, the objectives of this chapter are:

- Elaborating on two major emerging technological advancements and their implications in the construction industry: *How could these technologies change the built environment?*
- Finding the social impacts of implementing automated technologies on the human community: *What are the important considerations before implementing an automated system?*
- Correlating the impact of technological advancements with unemployment rate as a major economic factor: *What is the impact of automation on unemployment rate quantitatively?*

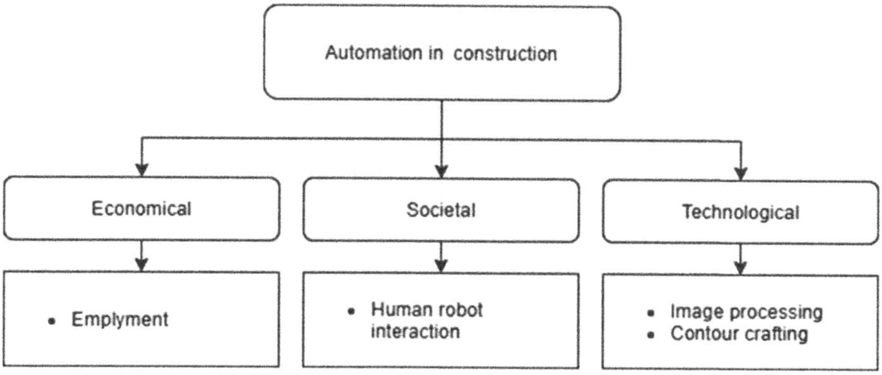

Fig. 5.1 Chapter's structure

5.2 Methodology

This chapter evaluates automation from three perspectives, namely, technological, societal and economical. Because of that, and more importantly due to the distinctive differences between these three topics, different suitable methods of research must have been selected. For the technological part, engineering design, numerical analysis and computer justifications had been used to develop an image processing unit. This unit can be used for separation of various particle types in recycled aggregate. Further on, upon evaluating the societal impact of automation, a systematic bibliometric analysis had been used to explore the opinions expressed by scholars. Jobs had always been replaced by new technologies throughout the history, and because of that, there's a colourful spectrum of literature available in that context. At last, where the economic evaluation becomes the subject of this study, a survey questionnaire had been designed to collect data from a target audience. The data was then used as an input in a statistical model called *Partial Least Squares Structural Equation Modelling* (PLS-SEM) to plot the correlation of robotics and unemployment.

5.3 Technological Aspect

5.3.1 Image Processing in Sustainable Urban Development

The need for affordable housing and population growth in metropolitans, together, dictates the operation of construction and demolition waste (C&D) recycling plants in a vicinity of smart cities. Following that and along with the implementation of sustainability principles, a proper investigation is required in order to assess the compatibility of an element as such within the urban environment, i.e. what are the components and how do they connect?

Touching down on the basic C&D supply chain, the recycling plant is being supplied with the C&D waste based on the local government plans for development and also average service life of buildings in a region. It is known that a metropolitan with about 6M people population generates roughly around 8M tonnes of C&D waste [5]. Recycling plants fundamentally operate following a set of procedures starting from sorting and then moving on to crushing, washing and screening, respectively. Noteworthy, each process by itself plays an important role in the final quality of the product which is called recycled aggregate. Throughout the whole process, a variety of heavy machineries are used. Crushers, as an example, typically consist of an electronic motor and a set of belts or a gear box which is currently being controlled manually by a human operator working on-site. Other machineries also require at least one operator to monitor the process. Now, when the concept of smart cities and IoT comes into place, it makes sense to shift the current manual and highly labour-intensive workflow into a system that can be operated from an operation room remotely. Such smart recycling plant can be seen as an element within the urban environment that not only internally is connected but also is capable of sending and receiving information to and from the future autonomous dump trucks.

Image processing as a high-end technology had also found its way in the C&D recycling [6]. This technology is currently at the laboratory scale in Australia with more studies needed to be conducted before moving on to a fully commercial scale. The preliminary studies had shown promising results on the utilization of image processing as part of an automatic sorting process which itself is a part of the whole recycling process. The study had been initiated at the first place when interviews with the concrete batching plants had revealed the lack of interest in using recycling aggregate in concrete. This is actually the driving demand part of the supply chain, i.e. recycled aggregate is widely being used as road base material and in some few instances used in concrete mix. The lack of interest in concrete contractors to use recycled aggregate in concrete is derived from the technical difficulties encountered. In fact, it had been stated that the water absorption of these aggregates is high in a way that negatively affects the development of the hydration reaction in concrete [7]. It can be said that this issue arises due to the presence of multiple particle types such as brick, old cement mortar, plastic, timber, etc., within a nominal batch of recycled aggregate, all with different properties adding uncertainties to the final output. In order to come up with a solution for this issue and ultimately encourage the concrete batching plants to see recycled aggregate as a reliable material, the sorting process needs to be modified. At the current stage, the sorting process is only summarized in separating steel rods by using strong magnets above the conveyor belt. Brick, old cement mortar, timber, and other particles remain in the batch. That indicates the need for fully automated sorting process that would be capable of, first, identifying the particle type and then commanding a robotic arm to pick and store each individual particle type in a designated area.

In an image processing system that was previously crafted, three main digital components were connected to a database in order to operate a sorting process as part of the C&D recycling process [8]. The three digital components are, namely, a camera, conveyor belt and robotic arm. The whole image processing and the sorting system are summarized as follows:

1. Recycled aggregates are poured on the conveyor belt which runs with a specified speed and a certain recurring delay.
2. A camera which is installed above the conveyor belt identifies the type of particles within the recycled aggregate feed on the conveyor belt.
3. The output from the camera contains information about the particle type, size and location on the belt based on a Cartesian coordination system.
4. The output from the camera, then, will be compiled as an input for the robotic arm. This means that the claws are notified about the location, type and size of the particles which indicated where to lift and discharge the particles.

The schematic protocol is as shown in Fig. 5.2.

At the moment the integration of the image processing camera and the compiler (control system) is solely developed, i.e. two other important connections between the robotic arm compiler and conveyor belt compiler had not been developed yet.

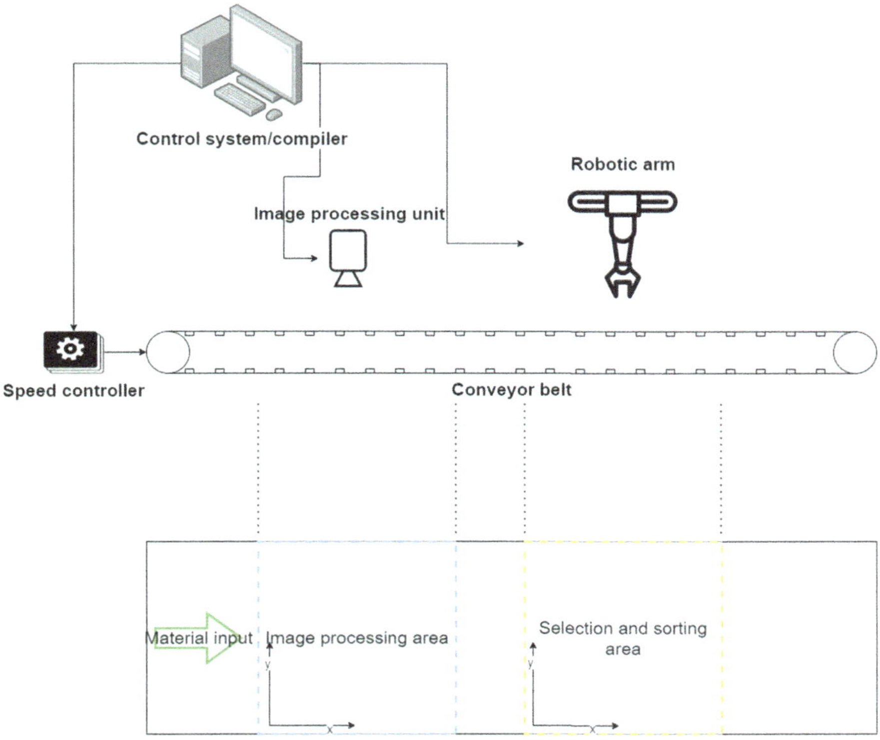

Fig. 5.2 The schematic workflow of the automated sorting system

5.3.2 Results and Discussions

The camera used for object detection in this study connects to a microcontroller which can be programed on Windows to input commands and also display outputs on the screen. The programing language used to control this system is C/C++. This integration operates in 3V–5V power inputs (Fig. 5.3).

The microcontroller has predefined libraries for the camera integration. The data received from the camera is compiled by the following C/C++ codes to output a matrix which can be used later as input data for the robotic arm. Since a large volume of data is being transferred between the module and the compiler, the speed of serial port needs to be adjusted accordingly. The program first identifies the blocks and counts the number of blocks subsequently using a FOR loop.

The accuracy of the system plays a crucial role in the successful commercialization process. In other words a need exists for the accuracy of the number of blocks identified and the size of the block reported by the machine. An array of eight elements/attributes had been produced each 1/60 of a second including the attributes such as the signature of the block (assigned particle name, e.g. brick, blue metal, plastic, etc.), x and y coordination of the block in the frame, width and height of the block, index number and age (which outputs the time-dependent sequence when the block appeared in the frame for the first time). A rough analysis on the output data had revealed that with a minor adjustment on the camera settings, the accuracy can be enhanced up to 87.5%. This largely depends on the quality of the recycled materials received as different batches may contain a variety of materials with different colours. One of the key important factors in the system adjustment is the colour code of the background. In case an operator decides to install the presented image processing system on a conveyor belt, the colour of the conveyor belt needs to be considered as an important factor during the calibration (Fig. 5.4).

Fig. 5.3 Integration of the microprocessor and the camera

```
#include <Pixy2.h>
// This is the main Pixy object
Pixy2 pixy;
VOID SETUP()
{
Serial.begin(115200);
Serial.print("Starting...\n");
pixy.init();
}

void loop()
{
int i;
// grab blocks!
pixy.ccc.getBlocks();
// If there are detect blocks, print them!
if (pixy.ccc.numBlocks)
{
Serial.print("Detected ");
Serial.println(pixy.ccc.numBlocks);
for (i=0; i<pixy.ccc.numBlocks; i++)
{
Serial.print(" block ");
Serial.print(i);
Serial.print(": ");
pixy.ccc.blocks[i].print();
}}}
```

Fig. 5.4 The code snippets for the image processing unit

Assigning signatures to the particles is another important aspect of this system. It requires a professional operator to calibrate the machine in an utmost efficient way in order to maximize the accuracy of the system in identifying the nature of the particles whether it would be brick, blue metal, plastic or others. A manual control system is available exactly for this purpose. Since the system runs on RGB (red, green, blue) processing, the intensity of each colour can be determined manually. In other words, the recycling batch needs a pre-assessment by the operator to distinguish the colour codes. For more accuracy, it is recommended to have a white LED installed beside the camera shedding light for more colour contrasting. Below an example of an unsuccessful object detection is presented in Fig. 5.5. The output data presented in Fig. 5.5 relates to a highly accurate processing where seven objects out of a total eight had been detected. The attributes recorded are presented in the same figure as well.

From the managerial perspective, an automatic sorting system as such is capable of enhancing the physical/mechanical properties of recycled concrete, while the final unit price of the recycled aggregate remains low. Concrete contractors, once the system had been utilized widely, shift to the use of recycled aggregate eagerly as

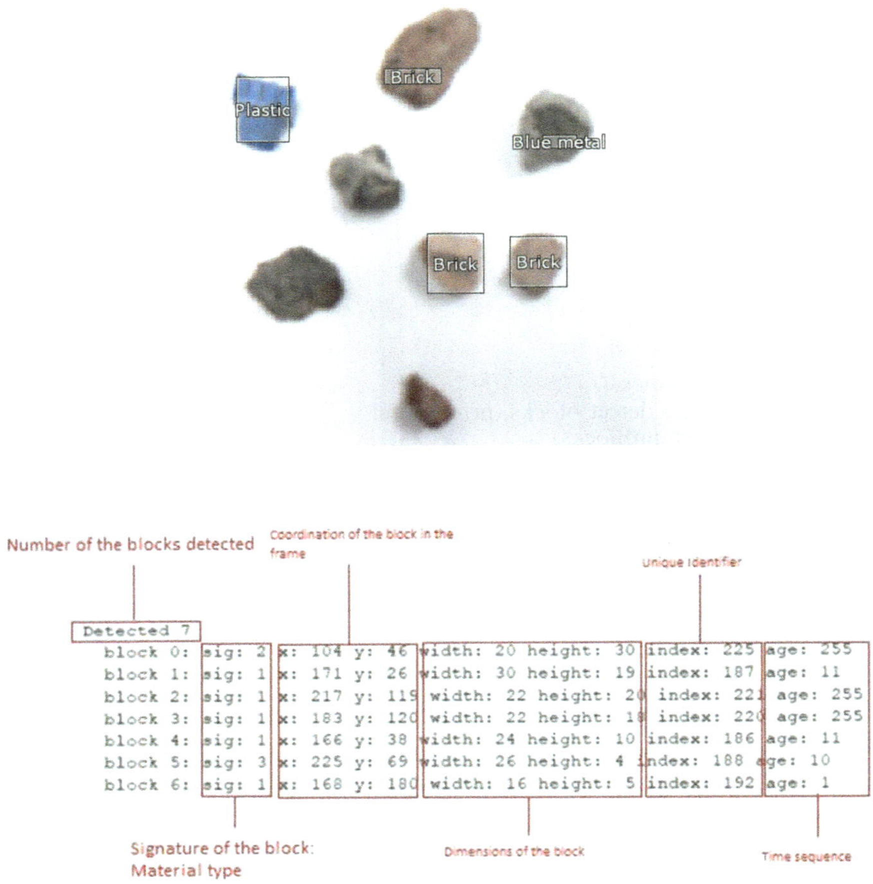

Fig. 5.5 Results of the object detection and the output matrix

opposed to the current hesitative preference of using natural aggregate. One can argue the economic impact of having such system operating in the market. Of course, it leads to loss of a few jobs as robotics often replaces old-fashioned jobs with new ones. Another economic aspect is the change in the supply and demand of recycled and natural aggregate. Once the eagerness to consume a higher quantity of recycled aggregate rises the demand for natural aggregate drops. Subsequently, as a positive result, the exploitation of natural resources and its terrifying consequences will be limited.

5.3.3 Contour Crafting

As an emerging technology, contour crafting compromises fast, accurate and safer construction technique. With an ever-increasing demand for housing and the boom in market prices particularly in megacities, one of the solutions is proven to be a much faster project delivery as such. Utilization of this technology in a wider scale allows simplification from the project management perspective, as well. Since the level of human interaction is lower in this method of construction, the manager can assure the stakeholders for a lower risk and a higher rate on investment gifted by the earlier delivery [9, 10].

The system is shaped by the integration of a computer program and a set of hardware devices. Several different programs and mechanical systems have been proposed with a slight advantage over the others; however, the foundation to all the system lays on four major factors: Cartesian coordination system, material cohesiveness/adhesiveness, pointer speed and flow speed. In construction, the tendency is towards the optimization of concrete contour crafting as the preferred extrusion material. It is yet to be investigated that what is the structural performance of buildings made by this technique. A critic that applies to this method is the lack of reinforcement in the segment's cross section which brings uncertainties for the tensile behaviour of the crafted elements [11].

5.4 Social Aspect

In terms of the change in human social behaviour at the presence of robots, it is necessary to divide the timespan into two distinguished periods as (a) transition mode and (b) fully automated cities. In the beginning of utilizing automated elements in smart cities, the human ambiguity in regard to social interaction with robots may cause inevitable hazards and delay in the processes. The reason why this might occur is the fact that learning to work with new objects usually takes a while for humans. In few words, in the transition period, as we can also see in the smart motorways, chaos is prone. Although there are some AI programing techniques that are equipped with self-teaching protocols, i.e. the robots are capable of constantly learning and adopting themselves with the environment.

Another significant impact that the utilization of robots may bring in the workforce is the loss of many jobs due to automation. It gets root from the economic aspect; however, the psychological effect that the loss of jobs has on the communities remains as an important social impact. Many people may fill reluctant after several years of education and experience in a particular field. What authorities need to consider before making a decision on whether or not to allow the presence of robots in the workforce is the number of humans who are going to be affected by this matter. As an alternative to overcome such social issues, the education sector must ensure that the generation to come is fully equipped with the skills required in a world where humans are competing with robots in terms of productivity.

F. Mayer in his article [12] urges the engineering designers to consider three different existing approaches in the design process, namely, techno-centric, socio-technical approach and human centred approaches. He had made the designer obliged to be aware of the "unquestionable fusion between automation technologies and society" and to use this as a way to strengthen the link between robots and humans. J. Cernetic et al. [13] emphasizes on the importance of engineering ethics, professional responsibility and public policy in this context. In the same article, it is mentioned that automation is mostly fearful for the group of people who poses a simple and repetitive job. Further on, the impact of intelligent automated systems is perceived differently by distinct group of people. On one side, management aims to achieve a better exploitation of resources, whereas the labours prefer to have easier job as a benefit of automation. Thus, it is of a high priority to comprehensively consider the implications of robotics before a proper implementation and use. Margulies and Zemanek in their article titled "Man's role in man-machine systems" [14] expressed that: "The troubles of our world are not caused by technology but by the way technology is being applied and used". Two of the main issues in the context of introducing a new technology had been mentioned by scholars: loss of motivation for work and, in some severe cases, loss of self-respect. Although these two issues had been only concerned by the way labours are affected, it can be implied that the close relatives of these groups of people will be affected as well. Once an automated technology is introduced, it mainly accomplishes the repetitive tasks which cause a significant change in the labour's work pace and rhythm. The operators of automated systems usually experience a long period of boredom followed by a sudden occurrence of concurrent alarms that causes enormous stress and mental workload, leading to the loss of self-respect in many cases.

Sheridan [15] in *Adapting automation to man, culture and society* quotes from Karl Marx that humans, in order to fulfil the inherent potentialities in itself, must engage in creative work. This is the reason for the being of automation as it escapes from doing repetitive tasks and at the same time the reason why automation degrades labour. He expands on the cultural norms and its causality on automation.

5.4.1 Cultural Norms

Historically, advanced societies record overzealous efforts to civilize the natives by pushing their sophisticated technologies on agrarian peoples. Such efforts often failed mainly because of primitive societies neither had the knowledge to maintain nor the required infrastructure to operate. "In such cases rubber-tired ox carts may function better and longer than the latest diesel truck". If a new technology/automation system is to succeed, the sake of technological transfer from one place to another and the economic opportunity brought by are insufficient. There are cultural norms in every society that must be understood by the developer, i.e. "The cultural

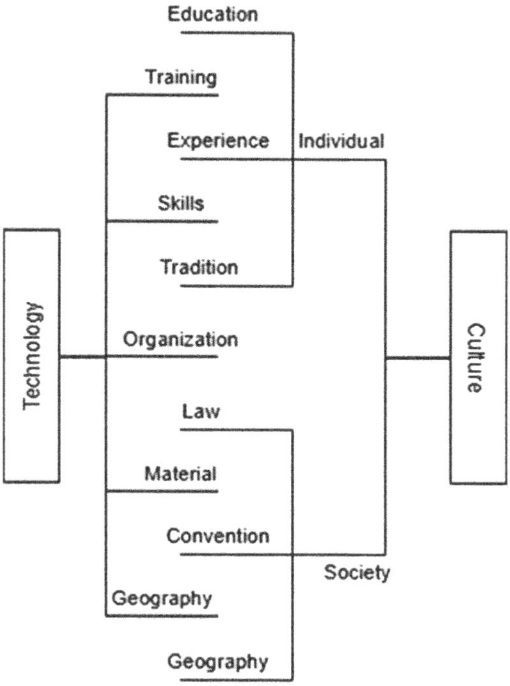

Fig. 5.6 An illustration for the factors influencing adoption of a new technology in a society

soil must be prepared by a proper philosophy of technological development". The society must precede a holistic perspective that is capable of receiving and synthesizing the technology into the culture, backed by a centralized management and flourished by individuals in the society that are driving the integrated technology with their cultural motives. Figure 5.6 conjugates the basic factors influencing the adoption of innovative technologies.

5.5 Economical Aspect

Productivity is the basic foundation in capitalism that defines the state of wealth in an organization/company. In a competitive environment, the entity which is capable of producing more goods faster and with a higher quality will have a higher chance of gaining higher market share and altogether necessitates the willingness to implement robotics in construction firms.

In the classical economic models, the presence of machines was being formulated in correlation with the number of human labours they take to operate, i.e. machines create jobs. However, in the modern post-Keynesian economics, the classical model is criticized based on the fact that robots are the reason for lower human employment. And the reality is in fact in that side as well because if one decides to think otherwise, the question "Why robots had been invented in the first place?" remains unanswered. It is sensible to state that robots help in brining productivity to a higher level and ultimately causing the GDP to grow, which is good. However, on the other side, the distribution of generated wealth and the loss of jobs require a reform in the monetary and fiscal policies in order to bridge the inequality gap between the social classes.

By definition, government monetary policy refers to actions and the regulations that a government puts into place in order to inject cash (supply of money) into the society. Since the amount of resources are limited and the power over money is the crucial key for any government, fiscal policies (collecting money from the society) are in place to work in conjunction with the monetary policies. Now, in transition towards an economy where productivity is reliant on automation (in our case within the construction industry), governors must be aware of the shift in the financial structure. Innovative companies usually introduce novel technologies that bring competition down in the pursuit of creating a monopoly. And that is a key factor in identifying the potential future leaders in the economy. Once a new technology emerges, the new company (which is the new money magnet) will have power over monetary and fiscal policies.

Employment, as another important factor in any economy, is subject to the level of automation in an industry. In the modern era, the major contradicting fact with what is known as post-Keynesian theory is the focus that is placed on the creation of jobs. In the post-Keynesian theory, it is claimed that machines are joint with the number of jobs that they create [16]. However, it is obvious in today's economy that as the functionality of robots are growing, with the integration of AI, many jobs are becoming reluctant. A clear example is the employment downfall during the COVID-19 crisis. A true impact of automation had appeared particularly in the education sector where the universities are in the complete lockdown; meanwhile they are fully functional satisfying the clients' (students') need and at the same time discontinuing many academic contracts as AI is taking over the place.

The major difference between the economy today and a century ago is the presence of robotics and artificial intelligence. Business entities, today, are not solely competing/cooperating with human being. Robots account for a large number of tasks being accomplished every day. And that causes loss/creation of jobs. One can view this phenomenon from the perspective that those who are utilizing automated robotic systems wish to accomplish daily repetitive tasks faster and more accurate. Otherwise, there will be no reason to benefit from robots. This ideology roots from the foundation where an individual believes that humans are reluctant against robots. Thus, he/she decides to replace human workforce with robots. On the other side, another group of people can believe that, even though robots are faster and more accurate, creating and operating these robots demand a new skill set and an emerging profession. Thus, it leads the creation of new jobs.

The situation however, inside the construction industry and by considering the universities' curriculum for construction related courses, leans more towards loss of jobs as a consequence of automation. Graduates from universities' construction degrees have less to no knowledge of computer science. And that indicates the fact that future construction experts cannot create and operate their own robots, leading a huge unemployment rate in the industry.

All the claims brought up so far in the previous sections remain as hypothesis unless it can be tested and verified by using *Partial Least Squares Structural Equation Modelling*, also known as PLS-SEM.

5.6 Conclusion

Technological advancement had always been an inevitable side of human livelihood. From centuries ago, by the industrial revolution and publicity of capitalism, management had defined the organization success via the means of fast and efficient exploitation of resources. These days, however, by the emergence of artificial intelligence, humans become reluctant from the workplace more than ever. Two of the emerging automated technologies in the construction industry had been explained in this chapter. Loss of jobs as one of the key areas in this context had been studied as well. Therefore, a precise focus must be regarded upon the implementation of any automated system as negligence often causes traumatizing social and economic consequences. It had been observed that continuous training and education in an automated built environment leads the organizational success.

References

1. J.M.D. Delgado et al., Robotics and automated systems in construction: understanding industry-specific challenges for adoption. J. Build. Eng. **26**, 100868 (2019). https://doi.org/10.1016/j.jobe.2019.100868
2. S.S. Kamaruddin, M.F. Mohammad, R. Mahbub, Barriers and impact of mechanisation and automation in construction to achieve better quality products. Procedia-Social Behav. Sci. **222**, 111–120 (2016). https://doi.org/10.1016/j.sbspro.2016.05.197
3. A. Oke, C. Aigbavboa, S. Mabena, Effects of automation on construction industry performance, in *Second International Conference on Mechanics, Materials and Structural Engineering (ICMMSE 2017)*, (Atlantis Press, 2017). https://doi.org/10.2991/icmmse-17.2017.61
4. A. Oke, C. Aigbavboa, S. Mabena, Drivers for the application of automation in construction activities, in *Commonwealth Association of Surveying and Land Economy (CASLE-Dar es Salaam)*, (2017), pp. 80–87
5. V. Tam, F. Sartipi, K.N. Le, Gaps between supply and demand of recycled aggregate: Sydney metropolitan case study. Presented at the CRIOCM 2018, 2018
6. A. Karji, A. Woldesenbet, S. Rokooei, Integration of augmented reality, building information modeling, and image processing in construction management: a content analysis, in *AEI 2017*, (2017), pp. 983–992
7. V.W.Y. Tam, A. Butera, K.N. Le, Carbon-conditioned recycled aggregates in concrete production. J. Clean. Prod. (2016). https://doi.org/10.1016/j.jclepro.2016.06.007
8. F. Sartipi, Automatic sorting of recycled aggregate using image processing and object detection. J. Constr. Mater. **1**, 3–3 (2020). https://doi.org/10.36756/JCM.v1.2.1

9. F. Bos, R. Wolfs, Z. Ahmed, T. Salet, Additive manufacturing of concrete in construction: potentials and challenges of 3D concrete printing. Virtual Phys. Prototyp. **11**(3), 209–225 (2016). https://doi.org/10.1080/17452759.2016.1209867
10. F. Craveiroa, J.P. Duartec, H. Bartoloa, P.J. Bartolod, Additive manufacturing as an enabling technology for digital construction: a perspective on construction 4.0. Sustain. Dev. **4**, 6 (2019). https://doi.org/10.1016/j.autcon.2019.03.011
11. S.H. Ghaffar, J. Corker, M. Fan, Additive manufacturing technology and its implementation in construction as an eco-innovative solution. Autom. Constr. **93**, 1–11 (2018). https://doi.org/10.1016/j.autcon.2018.05.005
12. F. Mayer, Social impact of automation trends and issues: an human centred systems engineering perspective. IFAC Proc. Vol. **41**(2), 5251–5256 (2008). https://doi.org/10.3182/20080706-5-KR-1001.00881
13. J. Černetič, S. Strmčnik, D. Brandt, Revisiting the social impact of automation. IFAC Proc. Vol. **35**(1), 167–178 (2002). https://doi.org/10.3182/20020721-6-ES-1901.01648
14. F. Margulies, H. Zemanek, Man's role in man-machine systems. Automatica **19**(6), 677–683 (1983). https://doi.org/10.1016/0005-1098(83)90031-6
15. T.B. Sheridan, T. Vámos, S. Aida, Adapting automation to man, culture and society. Automatica **19**(6), 605–612 (1983). https://doi.org/10.1016/0005-1098(83)90024-9
16. M. Lavoie, *Post-Keynesian Economics: New Foundations* (Edward Elgar Publishing, 2014)

Chapter 6
Human-Drone Interaction (HDI): Opportunities and Considerations in Construction

Gilles Albeaino, Masoud Gheisari, and Raja R. A. Issa

Abstract The construction industry is witnessing significant growth in drone applications, ranging from site mapping and surveying and progress monitoring to safety monitoring and structure inspection and maintenance. Drones' recent technological advancements and successful integration with other technologies, along with their ability to accomplish tasks safely, quickly, and cost-effectively, have made them popular robots on construction jobsites. In the upcoming years, drones will be an integral component of construction teams, and these flying robots are envisioned to cooperate collectively or with other types of robots while closely interacting with jobsite workers and personnel. This chapter focuses on: providing a comprehensive overview of the different applications of drone technology in construction; understanding the human-drone interaction elements, research areas, and opportunities; as well as summarizing the interaction considerations while proposing a future research roadmap that ultimately paves the way to safely and more efficiently integrate drones on construction jobsites.

Keywords Drone applications in construction · Human-drone interaction in construction · Interaction modalities · Comfortable approach distance · Aerial robot trust · Human-drone communication · Anthropomorphic drones

---

G. Albeaino (✉) · M. Gheisari · R. R. A. Issa
Rinker School of Construction Management, University of Florida, Gainesville, FL, USA
e-mail: galbeaino@ufl.edu; masoud@ufl.edu; raymond-issa@ufl.edu

H. Jebelli et al. (eds.), *Automation and Robotics in the Architecture, Engineering, and Construction Industry*, https://doi.org/10.1007/978-3-030-77163-8_6

6.1 Introduction

Originally, drones or unmanned aerial systems were intended for military purposes. However, recent technical advancements in robotics, engineering, and aviation made these platforms more ubiquitous and inexpensive, paving the way for utilizing this technology for different types of civilian operations. Construction, in particular, emerged as one of the top industries for drone adoption within the last decade [1], with this technology being capable of accomplishing tasks more efficiently compared to traditional drone-less procedures. Applications benefitting from drone technology adoption in construction cover the entire project lifecycle, from preconstruction (e.g., site planning and evaluation) and construction (e.g., site mapping, earthwork volumetrics, safety monitoring, and inspection) to post-construction (e.g., structure and infrastructure inspection, post-disaster assessment) [1, 2]. With the use of drones on jobsites, hard-to-access locations have become easily reachable, and construction processes have become safer and more efficient [1, 3]. For example, drones have been effectively assisting safety managers in the monitoring and inspection of jobsite safety hazards and helping in documenting jobsite conditions and project progress through time [4–6]. They have also been relied upon to accurately perform structural inspections and assess thermal leaks within buildings [7, 8].

This trend in drone adoption is expected to increase even more, especially with the inefficient traditional processes and lagging productivity rates which recently forced the construction industry to integrate automation and robotics into current practices. Combined with other technologies, procedures (e.g., photogrammetry, big data, AI, machine learning predictive analytics, and wearable devices) and technical advances, drones have recently started to shift roles and have moved from passively following control commands to actively interacting and collaborating with humans and assisting them in different tasks and activities [9–13]. These aerial platforms are expected to permeate jobsites, collaboratively operate with other aerial robots, work as a team with ground robots (e.g., self-driving bulldozers and excavators, autonomous machines), and continuously interact with workers and personnel using more natural user interfaces and control modalities to assist them in their wide-ranging and challenging construction tasks. Such factors enable human-drone interactions to become an integral and inevitable part of construction environments.

Coping with the expected increase in the interaction and collaboration between humans and drones on construction jobsites, there is a necessity for developing an understanding of the different elements, roles, research areas, and opportunities associated with such interactions. In addition, it is of particular importance to review the considerations and challenges associated with drone usage in construction and develop a comprehensive research roadmap while prioritizing different areas warranting further investigation. While the advantages and potential applications of drones have been thoroughly explored in the literature [1, 3, 14], there is a need for investigating the considerations and opportunities facing human-drone interaction within the construction jobsite context. The goal of this chapter is to provide a thorough understanding of the different drone elements of future human-drone/robot teams and study the challenges and considerations associated with their interactions. These factors will eventually lead to this technology's successful and safe integration in construction. More specifically, this chapter first provides a detailed overview of the drones' current and future potential applications in construction while highlighting the advantages of using this technology compared to conventional methods. It also defines the different roles humans and drones could play while accomplishing various tasks and exhaustively reviews the emerging human-drone interaction research areas currently being explored in the human-robot and human-drone interaction research fields. The commonly encountered drone implementation challenges are then summarized, and a research roadmap is proposed consisting of different potential research areas that warrant further investigation. The contribution of this study is advancing the understanding of human-drone interaction within the construction industry context. Specifically, the knowledge acquired in this study will ultimately help in recognizing the industry needs as well as identifying and mitigating the challenges and risks associated with human-drone interactions on jobsites, factors that ensure a safe and efficient interaction in this setting.

6.2 Drone Applications in Construction

Researchers and practitioners have been increasingly interested in integrating drones in construction and developing procedures to overcome common deployment challenges that this technology encounters in dynamic, complex, and hazardous environments such as construction jobsites [1, 14, 15]. This increased integration of drones in construction stems from the technical advancements that this technology recently witnessed, combined with the devices' abilities to accomplish tasks safer, in less time and cost, while being capable of accessing difficult-to-reach areas [1, 3]. Drone-mediated applications extend throughout the entire construction lifecycle, with applications ranging from site surveying and mapping and site planning [16] to structure and infrastructure inspection [17], progress monitoring, and building maintenance [1–3]. The following subsections examine current and future potential drone application areas in construction within the contexts of the pre-construction, construction, and post-construction phases of a project.

6.2.1 Pre-construction

During the pre-construction phase of a project, drones have been useful for site evaluation and site planning applications.

6.2.1.1 Site Evaluation

Site evaluation mainly consists of collecting information to assess the site's suitability for its intended construction purpose. Compared to conventional site evaluation methods, which often rely on satellite imagery, tachymetric surveying, and laser scanning, drone-acquired visuals provide managers with spatial and temporal information that would allow them to perform site feasibility studies in a cost-effective and timely manner [18]. For example, drone technology has been relied upon to conduct feasibility assessments prior to proceeding with the renovation of a nineteenth-century heritage monument [19]. Study results showed that drones offer an all-inclusive and multidisciplinary solution that would help public and private entities better identify and revitalize sites that have the potential of returning the most value on investments.

6.2.1.2 Site Planning

Drone technology could also be advantageous for site planning applications, where drone-acquired visualizations would be used by site managers for construction scheduling, as well as for planning jobsite layouts, activities, and logistics. Conventional planning tasks are challenging, usually requiring managers to have a clear understanding of the construction drawings and layout, which might even necessitate them to frequently walk the jobsite and collect relevant information pertaining to the site conditions [2]. Even with such practices, schedule, activities, and layout planning could still be delayed due to unexpected situations or poor decision-making. These factors encourage the use of drone technology as a supplementary site visualization tool to improve the planning accuracy of jobsite layouts, activities, and logistics. For example, drone-acquired images were combined with augmented reality (AR) to visualize a simulated version of a jobsite's construction progress, which helped practitioners improve site planning efficiency by predetermining unanticipated planning-related problems and mitigating them ahead of time [20]. A more recent study leveraged the use of drone-acquired three-dimensional (3D) spatial information in real time to improve current hoist site mapping and layout planning methods [21]. The authors tested their method on a petrochemical plant tower hoisting and showed that this technique saves time and improves hoisting operations' safety and efficiency compared to traditional 2D-based planning methods.

6.2.2 Construction

In the construction phase, researchers and practitioners have been employing drones for mapping and earthwork volumetrics, construction progress monitoring, construction safety monitoring and inspection, security surveillance, aerial construction and material handling, as well as site communication tasks.

6.2.2.1 Mapping and Earthwork Volumetrics

Drone-mediated mapping and earthwork volumetrics applications consist of using drones to survey, map, and collect spatiotemporal information pertaining to specific jobsite areas necessitating earth movement operations. Conventional surveying and mapping methods are mainly accomplished using tachymetry-based methods, global navigation satellite systems (GNSSs), laser scanning, and light detection and ranging (LiDAR) devices [18]. However, the onsite use of these techniques encounters several challenges, such as the relatively long time and high cost associated with these methods in surveying complex jobsites, favoring the use of drones as a timely, cost-efficient, and effortless alternative [18]. As an application example, a recent study compared drone-acquired 3D mapping data to traditional tachymetric-based surveying techniques serving as ground truth measurements to assess drone utilization as a mobile mapping and surveying tool [22]. After evaluating drones' efficiency on three different field-test areas (i.e., landfill, road construction, and high-speed rail construction), results showed the successful applicability of this technology for surveying and mapping projects involving earthmoving operations. A more recent case study utilized drone-acquired photogrammetric 3D models to evaluate the accuracy of this surveying technique in measuring earthwork volume change before and after the partial excavation of a gravel stockpile. The authors compared the results with ones obtained from conventional, real-time kinematic (RTK) GNSS-based ground control point (GCP) surveying [18]. Their analysis revealed similarities in the results of both surveying methods, encouraging the use of drones for medium-size projects due to the platforms' abilities to accomplish earthwork surveying tasks more safely and cost-effectively.

6.2.2.2 Construction Progress Monitoring

During the construction phase, drones could also be deployed to track the project's progress with respect to time. Conventionally, supervisors used to either manually collect terrestrial images by walking the site or rely upon satellite imagery for surveillance. However, these conventional methods are ineffective, challenged by the large amount of time associated with manual data collection and the low-quality resolution of the acquired satellite images [2, 23]. Drone-mediated construction progress monitoring is associated with several advantages, including the drones' high operational flexibility, location independency, and high-resolution imagery [24]. These advantages offer construction practitioners the ability to capture high-quality and detailed visualizations in less time and at a closer distance from the jobsite, significantly improving the overall onsite progress monitoring efficiency. For example, Unger et al. utilized drone-mediated photogrammetry to monitor, over a period of 5 months, the progress of the work occurring in a building zone [24]. Study results showed that the derived drone-acquired 3D products are accurate enough that even minor changes were successfully identified, recommending the use of drones as a cost- and time-efficient alternative to conventional progress monitoring tasks. Very recently, Ibrahim and Golparvar-Fard combined building information modeling (BIM) with drones to automate the drone flight planning process and data collection and improve drone-mediated construction progress monitoring applications [6]. Their method consists of optimizing drone flight missions by pre-identifying, through simulated flight paths, the ideal conditions to ensure a safe, detailed, and efficient data collection for documenting ongoing construction activities. After implementing their approach on a construction project, the authors noted improvements in terms of the acquired data completeness and visual quality, advocating the usage of this technique for monitoring construction progress.

6.2.2.3 Construction Safety Monitoring and Inspection

Construction is ranked as one of the top sectors susceptible to injury accidents and fatalities [25]. Conventional construction safety monitoring and inspection tasks entail safety managers frequently walking jobsites, identifying unsafe behaviors and situations that would put workers' and practitioners' lives at risk, and mitigating them. Several risks are typically associated with these tasks, including: (1) safety hazards associated with managers trying to access difficult-to-inspect and hazardous locations, possibly affecting their overall construction safety monitoring and inspection performance; and (2) relatively low and insufficient number of safety managers available on the jobsites, which could negatively affect their overall ability to effectively detect and mitigate all hazards that workers are exposed to [4]. In this context, drone technology has proved to be an efficient safety monitoring assistant, capable of performing onsite inspection tasks quickly and accessing hard-to-reach locations in a timely manner. As an example, Gheisari et al. developed a fall hazard identification algorithm that relies on drone-acquired and photogrammetrically generated 3D models to automatically detect guardrails and openings, not in compliance with the established regulatory safety standards [26]. More recently, Martinez et al. explored the feasibility of integrating drones into current safety monitoring and inspection processes and presented the limitations that should be overcome for the success of such integration [4]. Through a high-rise building case study, the authors showed that the utilization of this technology improved safety hazard identification and assessment.

6.2.2.4 Future Drone Applications

Security Surveillance: The construction sector is a prime target for thieving activities. The costs of equipment theft on construction jobsites were estimated between \$0.3B and \$1B annually [27]. One of the future drone applications in construction will be to utilize these platforms for security surveillance tasks, protecting against stealing and trespassing. Currently implemented jobsite surveillance practices include the use of closed-circuit television (CCTV) video surveillance, stationary cameras, or even relying on security guards to protect the site and prevent any intrusion. In the near future, drones will be equipped with more efficient and long-lasting batteries that would allow this technology to constantly provide real-time feedback on the jobsite's status and condition during after-hours and overnight. The advantages of using drones compared to current methods include the significant reduction in costs, particularly on large construction projects, as well as the platforms' ability to access hard-to-reach areas quickly and safely. These advantages make drones well-suited for efficiently performing onsite security surveillance tasks in the imminent future. As an example, Bürkle et al. [28] developed a fully autonomous drone-ground control station system integrating multiple individual aerial platforms to improve the implementation and effectiveness of drone teams in complex tasks. Despite not assessing their system in real-world conditions and only being limited to a virtual reality (VR)-based site intrusion scenario, the authors suggested different drone swarm deployment applications and highlighted the abilities for drones to autonomously operate and detect any site trespassing or violation.

Aerial Construction and Material Handling: Drones are characterized by their location independency and operational flexibility, being capable of flying to almost any jobsite location. Such characteristics pave the way to apply this technology for aerial construction and material handling types of tasks, where these platforms would be controlled autonomously via specific algorithms to transport material and build high architecturally complex structures [2]. Doing so will ultimately eliminate the need for cranes or scaffolding typically required to conduct such types of construction operations. However, implementing drones for aerial construction and material handling tasks is associated with several challenges, including: (1) reduced tolerable payload weight and battery life, which restrict mounting the platform with heavy components; and (2) the efficiency of the mathematical algorithms in precisely controlling drone swarms during the construction process [29]. These factors would require multidisciplinary efforts to improve the efficiency of drones in terms of sensor design (e.g., high-quality, robust, powerful, and lightweight sensors), platform development (e.g., power supply sources), and programming (e.g., precise handling and guidance systems, simultaneous cooperation of a swarm of drones). In addition, more efficient construction designs and processes, as well as other material alternatives, must be explored to enhance the adaptation of drones for aerial construction and material handling applications. In this context, and although limited to a mockup structure, Lindsey et al. [30] succeeded in autonomously controlling a swarm of drones to perform structural elements pick up, transport, and assembly tasks. More recently, collaborative work between the Université Catholique de Louvain and the Massachusetts Institute of Technology explored the feasibility of using customized drones to construct masonry structures [31]. Preliminary laboratory testing showed promising results regarding the drones' capability of transporting materials and building heavy and architecturally complex structures, factors that warrant additional studies to evaluate, in real-world construction environments, the effectiveness of this technology.

Site Communication: The ability of drones to be equipped with onboard cameras and wireless audio/video transmitters makes them ideal communication tools on construction projects [32]. By using drones, jobsite managers will be able to instantaneously guide, provide feedback, and communicate with their workers, regardless of their location on the jobsites. Site communication is an essential measure of project completion success. In fact, effective communication among different project entities improves teamwork and collaboration and reduces delays and disputes, ultimately resulting in timely and cost-efficient project completion. As a proof-of-concept application, Schneegass et al. [10] mounted a tablet on a drone to investigate, through a user study, the feasibility of using free-floating aerial displays for a variety of communication-related applications. Participants' interview results highlighted the importance of using drone-mediated aerial displays, particularly when communicating with and providing information to groups of people.

6.2.3 Post-construction

During the post-construction phase of a project, drones have been primarily used for performing structure and infrastructure inspections, assessing structures in the post-disaster setting, as well as for marketing purposes.

6.2.3.1 Structure and Infrastructure Inspection

Structure and infrastructure inspection applications are primed for growth in drone technology adoption. In fact, structure and infrastructure inspections are the most common drone applications within the construction industry [1, 3]. These inspections mainly consist of using drones for assessing the condition and structural integrity of structures, including bridges, buildings, cultural monuments, roads, solar panels, dams, and retaining walls. Most traditionally used building and bridge inspection methods include under-bridge inspection vehicles and elevating platforms, scaffolding, as well as rope access [33]. However, these methods have several drawbacks, including: (1) the high equipment and indirect costs; (2) potential safety hazards particularly when accessing hard-to-reach areas and using the rope access and scaffolding methods; as well as (3) the time required for logistics (i.e., mobilization, inspection, and lane closure), factors that encourage the use of more innovative inspection alternatives in this setting [33]. In this context, drone technology emerged as an efficient inspection tool, capable of accessing difficult-to-access areas and minimizing cost, time, and safety hazards associated with conventional inspection tasks.

Structure and infrastructure inspections using drones have been widely explored in the literature. Duque et al. [34] explored drones' efficacy as a supplementary bridge inspection and damage quantification tool by developing a four-stage damage quantification protocol and testing it on a timber arch bridge. Comparative results with conventional actual onsite field measurements showed that the crack rust stain areas, thicknesses, and lengths quantified using the drone-based protocol were accurate, with minimal differences in values between both methods. More recently, Tomiczek et al. [35] relied upon a custom-designed drone specifically equipped with an upward-directed camera and obstacle avoidance (i.e., optical flow device, laser range finder) sensors and tested its effectiveness by inspecting eight different GNSS-deprived bridges. Despite not providing similar information compared to conventional under-bridge inspection vehicles, drone utilization was recommended for special and damage inspections and as a periodical alternative to conventional vehicles for inspecting bridges in good conditions.

Drones have also inspected buildings for various uses, including damage detection and quantification as well as thermal and energy assessments. For example, Roca et al. [8] conducted comparative analyses between a point cloud of a building facade generated using a drone equipped with a Red-Green-Blue-Depth (RGB-D) camera sensor and an equivalent one acquired using the conventional terrestrial laser scanner method. The authors demonstrated that the drone-generated point cloud compared well with traditional methods, recommending the use of this technology as a standalone building inspection alternative or a complementary solution for traditional inspection methods. The indoor environment of buildings, which is characterized as a challenging environment for drone navigation due to the absence of GNSS, was also inspected using aerial vehicles. A recent study leveraged the use of ultrasonic beacon systems, deep convolutional neural networks, and geo-tagging to autonomously operate drones in GNSS-denied environments and perform indoor building inspection tasks [36]. Case study results showed that concrete cracks were accurately localized and detected, advocating the usage of this technique in this setting. Drones were also equipped with thermal imaging devices to inspect buildings. For example, this technique was successfully implemented on the curtain walls of a building facade to detect potential thermal bridges, quantify their magnitudes, and compare them to standard temperature values [7]. Drones have also proved to be efficient in inspecting cultural monuments, roads, solar panels, dams, and retaining walls [1].

6.2.3.2 Post-disaster Assessment

Natural disasters such as hurricanes and earthquakes could have devastating effects on structures and infrastructures, requiring immediate and urgent damage evaluation and rehabilitation. The conditions of such environments could be extremely difficult to an extent where the inspectors' ability to access the affected area and assess the damages would be restricted. To overcome such limitation, inspectors typically used to either: (1) compromise by putting their lives at risk when accessing affected areas; or (2) rely on the low-quality satellite images that, despite being helpful in visualizing large-scale areas, fail to provide a complete and detailed analysis of the structures' conditions on an individual scale [37]. Drones' location independency advantage, combined with their ability to quickly and safely provide a bird's-eye view of the location of interest, make them ideal for emergency response missions, as well as for the inspection of structures and infrastructures damaged by natural events. For example, drones were successfully combined with ground robots to collaboratively and safely assess the damages that buildings, cultural heritage monuments, and artifacts endured in multiple earthquake events [38, 39]. Despite encountering some drone deployment challenges (e.g., magnetic interference caused by the presence of metal component indoors, limited drone accessibility to some locations, drone navigation in GPS-deprived or indoor locations), the results of both studies highlighted the importance of drones in accessing difficult-to-reach or dangerous environments and encouraged the use of this technology for post-disaster damage identification and assessment. These findings were also validated in a more recent post-tornado study, in which drone-acquired visuals proved superior to conventional lower-quality satellite images in the identification of specific building elements [37].

6.2.3.3 Marketing

After the completion of their projects, construction companies, real estate professionals, and architectural firms heavily rely on drones for marketing purposes. Compared to ground-based photography traditionally used for marketing, drones provide a unique bird's-eye view perspective of the location of interest. Project stakeholders utilize drone-captured images and videos to either publicize and promote their projects during customer presentations and proposals for future potential projects or persuade prospective clients to invest [2]. Drone imagery could also be of great marketing and presentation value when integrated with: (1) VR, AR, and augmented virtuality technologies to allow clients to better envision their projects using indoor and outdoor virtual walkthroughs; and (2) BIM by overlaying drone-acquired photos and videos on top of BIM-based 3D models [2].

6.3 Human-Drone Interaction: Definition

6.3.1 Definition

Human-drone interaction (HDI) is an emergent multidisciplinary research area deriving from the well-established fields of human-robot (HRI) and human-computer (HCI) interactions, focusing on designing and investigating the interaction dynamics between humans and drones [40–42]. While HRI and HDI are somehow related in the sense that drones can also be categorized as robots, multidisciplinary research combining researchers and practitioners from the engineering, computer science, and cognitive science domains showed that not all HRI findings are directly applicable to aerial platforms [43]. In fact, the emergence and uniqueness of the HDI research sub-field stem from: (1) the drones' recent technical (i.e., software and hardware) advancements; (2) their efficient and wide-ranging deployment in a variety of tasks; (3) their distinctive natures of flying; as well as (4) their improved capabilities to unpretentiously perceive, be perceived by, and interact with humans from elevated and farther distances compared to other platform types (i.e., ground robots) [42, 43]. Moreover, while HCI and HRI research has been ongoing for several decades, HDI has been only present for the last 10–12 years [41, 43]. HDI research is broad and spans multiple domains, ranging between exploring the interaction modalities and approach distances that humans would mostly feel comfortable with while interacting with drones, as well as investigating human-drone communication and the platforms' feedback to the users' actions. Other researchers are also developing drones with anthropomorphic features to make them more social-friendly, appealing, and trustworthy, contributing to an enhanced human-drone interaction.

6.3.2 Roles of Humans in Human-Drone Interaction

HDI research showed that humans could take on different roles based on the application and interaction contexts [44]. It should be noted that these roles could be either active or passive, depending on the humans' level of involvement and whether they are controlling or being controlled by drones. As defined in [44], humans could play the role of operators or active controllers. In such situations, the drone's role would be to passively follow operational commands transmitted remotely to the aerial platform via human-controlled ground stations [1]. Application examples include tasks in which drones fully depend on pilots' judgment and decision-making, such as flying drones manually for capturing visualizations or advertisement purposes. The level of human involvement in such a role is normally high. Humans could also act as supervisors while interacting with drones [44]. In some instances, humans' involvement in drone-mediated tasks would be moderate, limited to supervising the flight mission and ensuring a safe operation. As an example, humans act as active observers in fully autonomous types of flight missions, in which drones are being passively controlled using pre-defined flight paths and parameters. Finally, humans could also play the role of recipients [44]. Contrary to the two previously defined human roles, this interaction will mainly be driven by drones with minimal, if any, human involvement. Some task examples consist of drones approaching people during emergencies for safe evacuation and crowd control purposes [10] and remotely delivering medical supplies and packages to intended users [45]. Figure 6.1 illustrates some application examples representing the different human roles in HDI, as well as their corresponding levels of human involvement.

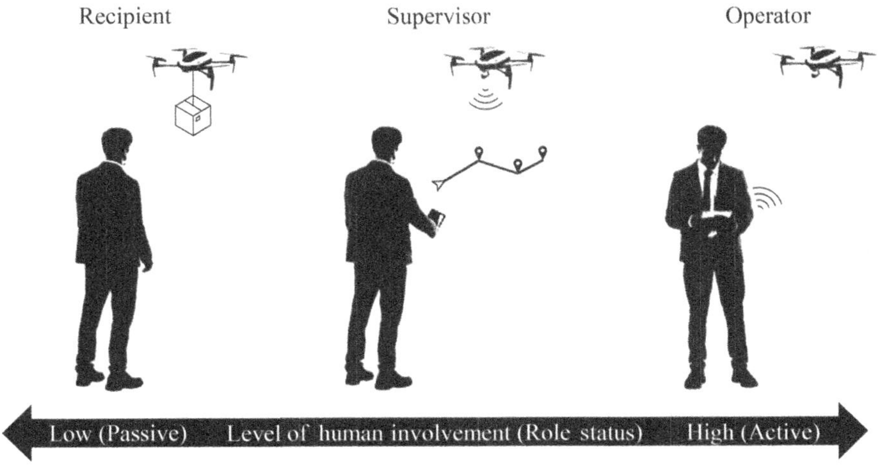

Fig. 6.1 Different human roles and levels of involvement in HDI

6.3.3 Roles of Drones in Human-Drone Interaction

Drones' efficient abilities to accomplish various tasks while interacting with humans enable them to take on different roles in HDI research [41]. First, they could act as human social companions. For example, drones were programmed to simultaneously recognize and follow performing athletes while providing them with external drone-acquired imagery, improving their motivation and performance [46]. They were also used to accompany runners to make jogging and physical exertion activities more engaging [13, 47]. Research has also shown that drones could assist humans by facilitating personal and general operations. Acting as personal assistants, drones have been relied upon to assist people with disabilities (i.e., blind joggers and visually impaired individuals) to better navigate environments [11, 12]. Examples of drone-assisted general operations include: (1) crowd control during emergencies and search and rescue operations, whereby drones would approach people and guide them to more secure areas [10]; and (2) sustaining a clean environment, in which drones would encourage people to throw their waste in the closest bins identified using aerial platforms [9]. Drones could also play the role of recipients by simply following humans' control commands. This role is particularly important and the most commonly used one in the construction industry, whereby drone-mediated applications have been proven efficient, allowing safe task accomplishment at reduced time and costs compared to conventional drone-less procedures [1].

6.4 Research Areas of Human-Drone Interaction

The following section exhaustively reviews the emerging areas of research that academics and practitioners are currently exploring in the field of HDI. These areas include (1) the human-drone interaction modalities, (2) human-drone comfortable approach distance, as well as (3) human-drone trust.

6.4.1 Interaction Modalities

Traditional interaction between humans and drones is typically accomplished one way, with drones controlled by humans via a remote controller and ground control station. While such means of interaction has the advantages of maintaining a low-latency and precise drone-operator communication, remotely controlling drones is limited by its unintuitive nature, requiring high levels of pilot training to operate the platform safely and assertively [14]. These factors, combined with the technical advancements that drones recently witnessed, motivated researchers to investigate other innovative and more natural human-drone interaction modalities or natural user interfaces (NUIs), including interactions based on gesture, speech, gaze, brain control, and touch (see Fig. 6.2). Each of these natural user interfaces has distinct characteristics, advantages, disadvantages, and applications that are further

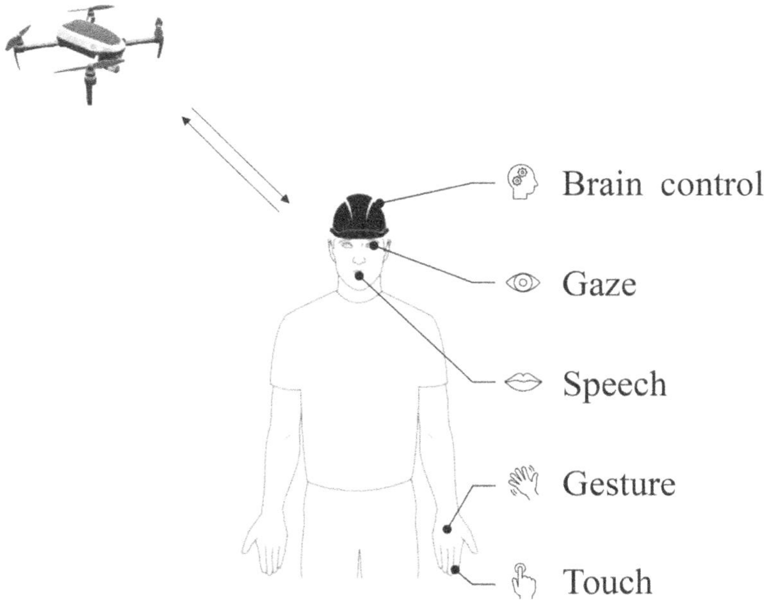

Fig. 6.2 Natural human-drone interaction modalities

Table 6.1 Summary of human-drone interaction modalities and corresponding characteristics

Human-drone interaction modalities	Characteristics					
	Latency	Control accuracy	Intuitive	Required pilot training time	Hands-free	Restricted to collocated spaces
Gesture	Moderate	Low	Yes	Short	No	Yes
Speech	High	Low	Yes	Short	Yes	Yes
Gaze	High	Low	Yes	Short	Yes	No
Brain control	High	Low	Yes	Long	Yes	No
Touch	Low	High	Yes	Short	No	Yes
Traditional (wireless telemetry)	Low	High	No	Long	No	No

discussed in the following subsections and summarized in Table 6.1. It is noteworthy to mention that such natural modalities are not mutually exclusive, and humans could rely on a multi-modal type of interaction that combines multiple modalities to compensate for each other's limitations [48].

6.4.1.1 Gesture

Gesture-based interaction consists of humans or operators relying on their body part movements (e.g., head, hands) to direct and communicate their intent to the drones to accomplish different flight tasks and maneuvers. Research showed that gesture-based interaction modality has several advantages compared to conventional, remotely controlled interaction. These benefits include: (1) enhancing the intuitiveness of the communication between humans and drones, a factor that may simplify drone operation and reduce the associated drone pilot training time; (2) wide compatibility with the majority of the commercially available aerial platforms that are already equipped with the necessary sensors (i.e., cameras) needed for establishing such human-drone communication; and (3) offering an interaction that, in contrast to remote controlling, does not necessitate the use of any additional devices except the ones mounted onboard the platforms to navigate and communicate with the drones [49, 50]. Conversely, the relatively low drone control accuracy and moderately latent gesture recognition and feedback provision limit the utilization of gesture-based interaction for specific construction applications and environments (i.e., collocated) compared to traditional remote drone control [51, 52].

Several researchers have explored gesture-based human-drone interaction for a variety of tasks and contexts. For example, Cauchard et al. performed a Wizard of Oz (WoZ)-based user-centered study to explore natural interaction techniques humans rely on to communicate with drones [48]. Results showed that out of 216 distinct interactions performed by the study's 19 participants, gesture was the most commonly used human-drone interaction modality, whether standalone or in combination (i.e., multi-modal interaction) with speech-based interaction. The same study was repeated in China and showed that cultural differences also affect how humans interact with drones, with Chinese participants relying mostly on gestures and more on multi-modal (i.e., using multiple modalities) types of interactions compared to their US peers [53]. A more recent study presented a drone-embedded visual interface that users can naturally interact with using gestures and evaluated its effectiveness in a drone-guided tour application [50]. The authors advocated the usage of this system to improve human-drone interaction intuitiveness and demonstrated its reliability, simplicity (i.e., necessitating minimal training), and responsiveness (i.e., being capable of reacting in real time to users' gestures).

6.4.1.2 Speech

Speech interaction consists of utilizing a combination of sounds and articulation to communicate and exchange information. Such type of interaction is straightforward and hands-free, only requiring drones to recognize human-instructed sets of voice commands and react accordingly while allowing users to channel their focus on commanding the drone itself rather than the ground control station [54]. The usage of such modality is associated with an increase in human-drone interaction intuitiveness which, similar to the gesture-based interaction, lessens the complexity of drone control as well as pilot training time [44, 54]. However, drones' voice recognition capability and latency might hinder speech adoption as a human-drone interaction modality, especially in complex and noisy environments such as construction sites, limiting the usage of such modality in non-collocated types of environments. For example, drones could be challenged by their ability to: (1) recognize, react to, and distinguish drone-targeting voice commands from other typical onsite worker communication; and (2) identify voice commands from farther distances, which might be problematic especially with the wide onsite usage of the noisy propeller-based rotary-type drones in construction [55]. While the limitation of identifying voice commands from farther distances may restrict the application of speech-based interaction to relatively small spaces, it can be overcome through voice command decoding using ground control stations [54].

Despite speech being the most predominant and instinctive form of human-human interaction, researchers and practitioners have not yet thoroughly explored this noninvasive modality as a means of interaction between drones and humans. In the previously described studies exploring how humans would interact naturally with drones [48] and the effect of cultural differences on such interactions [53] (see Sect. 6.4.1.1 "Gesture"), participants' results showed that speech was used by both Americans (38%) and Chinese (58%) to interact with the platforms. Another study proposed a drone-based software framework that enables users to interact with aerial platforms using different interaction modalities (i.e., speech, gestures, body position, visual markers, and multi-modal) depending on their intended applications [54]. Of particular importance was the speech interaction experiment, which consisted of the user choosing from a set of pre-defined commands, using speech to communicate this command with the ground control station, which in turn decodes the voice command before sending it to the aerial platform. Participants supported the usage of speech as an interaction modality, especially with the provision of feedback through voice synthesizers to confirm the proper receipt of the users' voice commands, and reported an analogy between their communications with drones and those with pets during training. The provision of feedback to confirm the proper receipt of users' commands as well as the addition of clearer and more flexible voice commands constituted some of the design implications of another WoZ-based study exploring multi-modal human-drone interaction in a search and rescue VR environment [56].

6.4.1.3 Gaze

Recent advances in eye-tracking technology and computer vision have paved the way to integrate gaze as another noninvasive and intuitive human-robot interaction modality. Gaze-based human-drone interaction mainly involves tracing current pointing directions of humans' eyes using eye-tracking devices located either in the environment or as head mounts on the human's body to control and navigate drones [57]. Gaze interaction could be valuable when combined with other modalities to enhance the interaction experience in general and reduce, for example, the ambiguity of referential expressions [58]. Another advantage of gaze is its ability to offer humans a direct, intuitive, and hands-free drone control and interaction alternative that is of particular relevance for people who heavily rely on their hands during work (e.g., construction workers) or others with mobility impairment (e.g., paraplegia) [59, 60]. The intuitive and hands-free nature of the gaze modality could also be beneficial for non-experienced drone pilots, allowing them to operate and control their drones safely and assertively [60]. On the other hand, gaze interaction requires the user to constantly look at the drone's flying direction, a limiting factor that might make drone control stressful and difficult, especially when the pilot is positioned in a different direction with respect to the drones or when the user looks away from the location the drone is planned to go to [59].

In terms of studies exploring gaze as an interaction modality, Hansen et al. presented four distinct gaze- and arrow key (keyboard)-based drone control modes to explore which of these options would be best to control each of the four degrees of freedom (i.e., speed, rotation, translation, altitude control) used for drone navigation in the 3D space [59]. Results showed that users were able to rely on gaze as a modality to interact with drones without necessitating much training and considered gaze control for rotation and speed combined with a keyboard for translation and altitude as the most reliable control mode. More recently, Yuan et al. relied upon a camera-, gaze tracker-, and inertial measurement unit (IMU)-equipped eye-tracking glasses to assist humans with drone navigation tasks [60]. Using these glasses, the authors proposed an approach to detect the user's location relative to the drone and navigate the aerial platform by identifying the gaze direction through user head orientation tracking.

6.4.1.4 Brain Control

Controlling drones using brain signals has recently emerged as a new human-drone interaction modality due to machine learning advancements. In such type of modality, users would be equipped with noninvasive brain-computer type of interfaces (e.g., electroencephalogram or EEG) to detect, via electrodes mounted on the users' scalp, the electrical brain activity prior to applying machine learning-based algorithms to decode these brain signals and translate them into precise drone control commands [61]. Despite not being exhaustively studied in the HRI and HDI research, the brain control modality offers a hands-free interaction that would allow both healthy and physically impaired people, hands-busy workers, and unexperienced pilots to interact with drones in a more efficient manner [62]. Another unique advantage over other interaction methods is that brain-controlled drones could adapt their behaviors based on the pilots' mental states and workload levels to ensure a better and safer operation [44]. For example, a drone might be programmed to only operate under acceptable pilot workload levels; otherwise, it will not be user responsive. Some of the reported challenges of EEG-based brain-controlled computer interfaces mainly encompass: (1) the user's ability to remain focused and maintain high concentration levels; (2) the extensive user training time to allow computer recognition of precise brain signals; and (3) the relatively low control accuracy and slow drone/system response to the users' commands [44, 61, 62].

Studies exploring the adoption of brain-controlled computer interfaces as a means for humans to control drones in the 3D space are very recent, dating back less than a decade [63]. LaFleur et al. equipped users with an EEG-based brain-computer interface and asked them to maneuver an aerial platform through a series of foam-made circular targets to study the potential of relying on humans' brain signals for drone control [63]. Results showed the users' ability to perform accurate drone maneuvering tasks efficiently and timely, encouraging the adoption of brain control as a human-done interaction modality in this setting. Moreover, Kosmyna et al. proposed a co-learning training protocol to reduce the training time associated with the use of operant conditioning (OC) and other machine learning-based protocols required for brain signals' computer recognition [61]. Comparative analyses revealed that their approach resulted in a similar to slightly lower performance within shorter initial training periods when compared to the OC protocol. These analyses resulted in their interface being specifically recommended for recreational type of activities.

6.4.1.5 Touch

Faced by the ubiquitous usage of drones and their increased associated safety risks, several researchers and manufacturers attempted to develop safer and more efficient platform designs. Improvements in designs consisted of software (e.g., failsafe autonomous algorithms) or hardware (e.g., safe-to-touch drones and addition of propeller guards and skin-protective cages) solutions [64, 65]. The hardware solutions, in particular, paved the way for the adoption of touch as a new, intuitive, and simple human-drone interaction modality [66]. While touch-based interactions offer the advantage of integrating drones into a wide variety of applications that necessitate contact between humans and drones (e.g., sports and VR), this interaction modality requires users to mount the platforms with additional safety components (e.g., propeller guards and cages) and remain in close proximity to the drones, which limits this modality's adoption only to collocated environments [44, 65, 67].

Abtahi et al. [66] relied upon two versions of a quadcopter: the original non-modified commercial design, and another customized design featuring a safe-to-touch lightweight wooden frame and mesh around the propeller blades, to assess whether participants will intuitively rely on touch while interacting with the safe-to-touch quadcopter. After performing a WoZ-based between-subjects study, the authors demonstrated that more than half of the participants touched the customized platform, and most of them felt safe interacting with the safe-to-touch drone. They also showed that interacting with the safe-to-touch drone significantly reduced the mental demand compared to the original drone version. More recently, a safe-to-touch drone equipped with a lightweight, protective cage was used as a haptic feedback device to interact with users in a VR-based boutique shopping experience [65]. In the VR environment, three different tasks were performed by the users using the safe-to-touch drone: (1) touching and feeling the textures of some clothing items using the fabric that was attached to the drone's protective cage; (2) holding up a clothes hanger that was manually installed on the platform; and (3) picking up the drone itself as a virtual shoebox prior to putting both the hanger and shoebox in the cart. Despite the propellers' wind and loud noise, most of the users enjoyed the experience, felt safe while interacting with the safe-to-touch drone, and supported drones' usage as haptic devices in this context.

6.4.2 Interaction Distance

The interaction distance is another fundamental research area in the HRI and HDI fields, mainly consisting of studying the factors that affect the comfortable approach distance between humans and other interactants, including robots and drones. The majority of the studies investigating human-drone comfortable approach distances base their findings on Hall's definition of the interaction spaces, originally intended for human-human interactions [68]. Hall defines four different spaces along with their corresponding distances: intimate [<0.46 m (1.5 ft.)], personal [0.46–1.2 m (1.5–4 ft.)], social [1.2–3.7 m (4–12 ft.)], and public [>3.7 m (12 ft.)] (see Fig. 6.3).

Duncan and Murphy queried the literature to identify the factors affecting the comfortable distance in human-human and human-drone types of interactions and showed that the approach distance is mainly influenced by environmental, personal, and agent-related conditions [69]. In addition, the study exploring the natural human-drone interaction modalities that was described in the previous section showed that American users felt safe interacting with the drone, with 84% of them

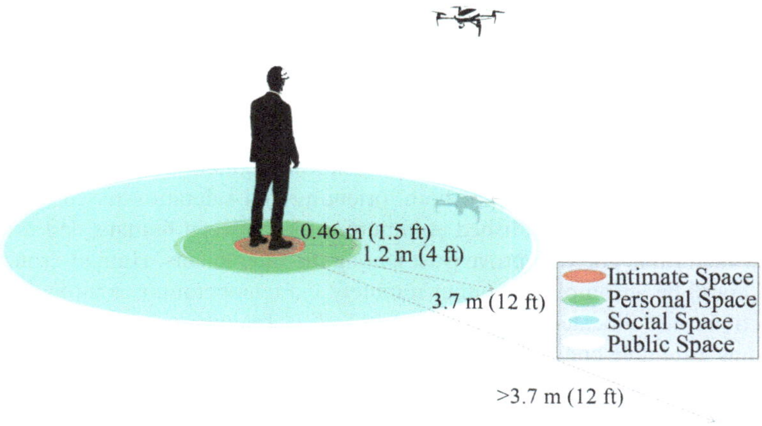

Fig. 6.3 Zone spaces along with their distances

allowing platform operation either within their intimate (37%) or personal spaces (47%) [48]. The other 16% controlled their platform within their public space due to the noise and wind of the drone propellers. On the other hand, the replicated Chinese version of this study demonstrated the effect of cultural differences on the humans' interaction distances with drones [53]. In fact, Chinese participants felt even more safe and comfortable interacting with the aerial platforms, with 88% of the users operating the platform within their intimate (50%) or personal (38%) spaces, and 12% controlled it within the social (6%) and public (6%) spaces. More recently, Acharya et al. conducted a within-subject analysis to evaluate, in an experiment room equipped with motion detectors, whether humans will exhibit different comfort levels while interacting with drones compared to ground robots operating at similar heights [70]. Contrary to the available human-ground robot and human-human literature, their study results showed that humans may distance themselves differently while interacting with these two robot types. This study highlights the distinctive drone characteristics, differentiating the well-established HRI research field from the emerging HDI.

6.4.3 Human-Drone Trust

Modern robotics and engineering advances changed humans' perception and understanding of the robots' roles in day-to-day tasks. Recently, humans started to perceive robots as active team members and social companions, rather than being passive, simply following users' operational commands. In this context, establishing trust between humans and robots is essential for a successful interaction and to increase drone adoption for a variety of different applications. Studies show that trust between humans and drones can be improved using several factors, one of which consists of enhancing both humans' and drones' abilities to perceive, acknowledge, and naturally interact with each other [44]. As an example, Jensen et al. investigated orienting and salutation-based (i.e., nod, toss, and waggle) drone gestures that humans would rather prefer for the aerial platforms to acknowledge their presence [71]. After evaluating the separate and combined contributions of these gestures, study results showed that acknowledgment is best communicated with the use of a combination of both, orienting and salutation-based drone gestures. Ensuring a well-established communication between humans and drones is another factor that could improve trust among the interactants. Human-drone interaction could be accomplished either remotely via teleoperation or through natural interactions in collocated environments [43, 72]. Teleoperation has already been proven as a reliable and efficient drone control method, especially with the recent use of low-latency networks such as 5G which pave the way to adopt drone technology for a wide variety of additional tasks, including ones necessitating beyond visual line of sight (BVLOS) operations [44]. However, this control type might not be as efficient and feasible in collocated spaces or when drones are being used as social companions [72]. In such environments, these conditions require more natural (e.g., gesture, speech, gaze, brain control, and touch) types of human-drone interactions. As an example, Ng and Sharlin proposed a socially motivated natural human-drone interaction technique using bird-based body gestures as a replacement for traditional remote controlling methods in collocated types of environments [72]. Despite being limited by the low number of subjects, the authors conducted two exploratory design critiques and showed that participants were receptive to most of the suggested gestures, as evidenced by their final emotional and social attachment to the drone. Human-drone trust could also be enhanced by designing anthropomorphic and human-friendly types of drones, characteristics that ultimately improve the social, emotional, and intuitive aspects of the overall interaction [44]. For example, Cauchard et al. evaluated participants' abilities to recognize a drone's emotional state by generating three different drone personality models, each characterized by distinct flight control parameters (i.e., speed, reaction time, altitude) and special movements [73]. By observing the drone's movement and reaction to the control commands, users accurately identified the platform's behavior and emotional state. A more recent study collected information from participants pertaining to the features and characteristics that an aerial platform must have to make it more social and designed a prototype accordingly [42]. The final design incorporated anthropomorphic characteristics and safety features and consisted of a social shape (i.e., oval-shaped protective cage), a face (i.e., Android tablet), as well as voice (i.e., text-to-speech-based greeting voice) as an additional interaction channel. Study results showed that users felt more comfortable interacting with the social drone, operating it at closer distances when compared to another nonsocial prototype.

6.5 Challenges and Future Research Roadmap

The adoption of drones in construction is associated with several barriers limiting this technology's full integration in the domain [1, 15]. The barriers that researchers and practitioners face in the HDI field were similar with respect to the ones encountered when deploying drones in general [44], including in the construction domain. For this purpose, this section aims at exhaustively reviewing the commonly encountered drone implementation challenges and provides construction researchers and practitioners with a research roadmap consisting of different potential research areas warranting further investigation. Figure 6.4 summarizes each of these research areas, which were identified based on the challenges (see Table 6.2) found in the literature [1, 3, 5, 14, 15] and grouped under four distinct categories: drone, human, regulations, and environment.

Drone: The literature shows that construction researchers encounter technical (i.e., software and hardware) challenges while using drones on jobsites [1, 15]. Examples of such challenges include: (1) loss of GNSS signal, potentially causing inaccuracies in drone positioning and navigational deviances from the originally pre-determined autonomous flight paths; (2) loss of radio communication between

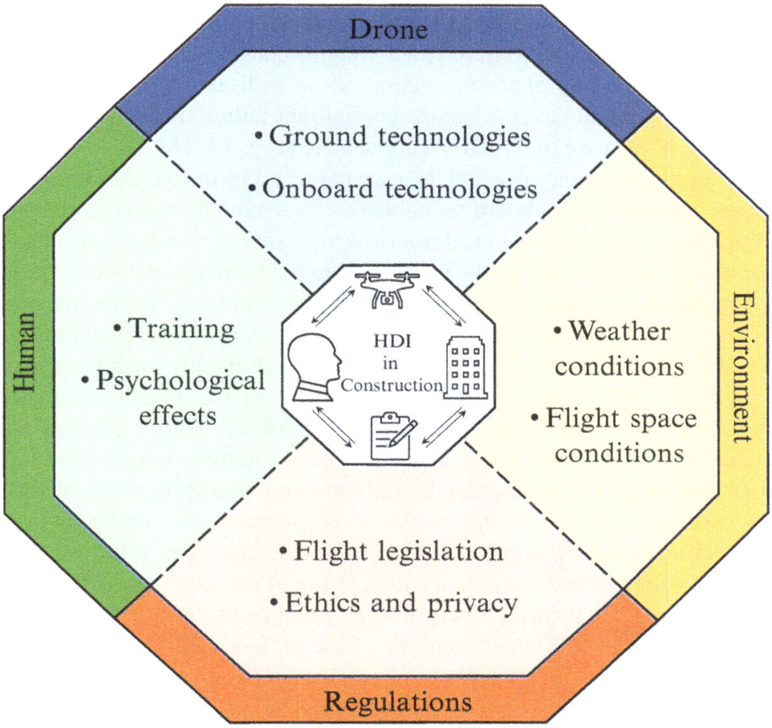

Fig. 6.4 Future potential HDI in construction research areas

Table 6.2 Summary of HDI challenges in construction

Construction-related HDI challenges	Citation examples
Radio communication or GNSS signal loss	[15, 74, 75]
Signal loss or interference	[4, 14, 26, 76, 77]
Complexity of user interface and ground control station	[5, 15, 78]
Limited payload weight	[22, 74, 79]
Limited hardware performance (e.g., short battery life, low-quality visual sensors)	[4, 5, 14, 26, 77, 78, 80, 81]
Insufficient pilot skills, training, and expertise	[5, 14, 77, 81, 82]
Negative worker psychophysiological effects	[4, 14, 26, 76, 77, 82]
Restricted flight regulations	[4, 14, 26, 78, 81]
Ethics- and privacy-related concerns	[5, 14, 78, 83]
Inclement weather conditions and unexpected natural events	[4, 76–78, 80, 81]
Presence of jobsite obstacles and crowded surroundings	[4, 5, 78, 81, 84]

the ground control station and the drone itself, eliminating any sort of ground (i.e., human) control over the aerial platform; (3) drone signal interferences caused by the onsite presence of conductive magnetic materials (e.g., electronic devices, underground powerlines) and structural components (e.g., structural steel framing, metal roofs), endangering the safety of both drone and nearby people; (4) complexity of the user interface and ground control station, which could create drone operational challenges; (5) increased drone payload weight, causing significant reduction in battery performance and platform endurance; as well as (6) payload component failure such as onboard battery damage and inflight failure of the brushless motors or propellers, in the case of a rotary-wing vehicle [1, 5, 14, 15].

Future multidisciplinary research is warranted, and further studies must focus on the drones' ground and onboard technologies to overcome these aforementioned challenges. Research on ground technologies must investigate the means and methods that would ideally simplify the ground control station's user interface complexity and operation, as highlighted in multiple studies exploring drone integration in construction [5, 15, 78]. Less complex drone user interface and operation prevent additional drone-associated risks from occurring and already hazardous and challenging jobsite safety conditions from exacerbating, ultimately enhancing this technology's adoption in construction. In addition, ground control stations must be capable of: (1) incorporating fast and efficient 5G human-drone communication networks, which ensure a reliable inflight interaction and increase the deployment of drones for BVLOS types of operations; and (2) decoding user's interaction input (e.g., speech and body gestures) to efficiently translate them into drone actions, allowing for a smooth and low-latency integration of different types of natural HDI modalities. Currently, there is a very limited number of commercial open-source drone systems capable of integrating any of the previously discussed natural HDI modalities [44]. In construction, deploying drones in BVLOS operations using efficient communication networks is of vital importance, particularly for heavy construction and industrial types of projects. In such situations, the numbers of workers and managers might not be sufficient to cover the entire project area, and drones could be deployed BVLOS to, for example, deliver tools to and communicate with workers, aerially construct buildings, and collect aerial visuals of an offshore structure.

In addition, research on the onboard technologies should focus on exploring the drones' physical characteristics such as platform type (e.g., vertical takeoff-landing rotary-wing, fixed-wing, fixed-wing hybrid) and form factor, as well as optimizing the drones' hardware components, which mainly consist of the platform's body components (e.g., batteries, motors, propellers, electronic speed controllers), navigation and communication systems (e.g., electronic speed controllers, power regulators, flight controllers, telemetry modules, GNSSs, radio controller receivers, and transmitters), as well as data collection systems (e.g., RGB, thermal, or multispectral cameras, LiDARs, digital video transmission devices, first-person view monitors). On jobsites, improving such hardware components will improve drones' robustness and performance and ensure a safe and reliable deployment, particularly in areas susceptible to signal interferences. Of particular relevance to the construction field is that future research should also explore improving both ground and onboard drone technologies to ideally enhance the usage of natural interaction modalities in non-collocated and exterior types of environments.

Human: To successfully integrate drones into the construction industry, humans must rely on more natural user interfaces (e.g., gesture, speech, gaze, brain control) and use very specific control commands to interact with the aerial platforms. This process typically requires long training periods and skills for the users to control drones safely and efficiently. The lack of skills, training, and expertise among pilots could negatively affect their judgment and decision-making, making drone operation a challenging task, especially in the dynamic and complex construction environments [5, 14]. Drone usage could also affect construction personnel on jobsites. In fact, flying drones on site could be associated with negative psychological effects (e.g., personal space invasion, increased stress, reduced levels of situational awareness) on workers, potentially distracting them from their original task and exposing them to hazardous situations [4, 14].

Based on the above, future research must focus on developing pilot training programs to train both novice and expert users on how to navigate drones through the newly emerged natural interaction modalities and flying styles. Indeed, such training needs to be performed in controlled and safe environments (e.g., experimental rooms, VR/AR-based simulations) to avoid any poor judgment that, if otherwise occurred on site, could potentially result in real-world hazardous situations. In addition, studies must explore the performance and psychological effects associated with the usage of drones among workers. Such influential effects could include mental workload, stress, comfortable approach and interaction distances, as well as situational awareness. For example, roofing workers might feel less comfortable and exhibit higher levels of mental workload and stress while interacting with drones compared to other onsite workers and humans. Exploring such effects on construction workers is vital to assess whether onsite drone deployment effectively improves the safety aspect of the construction processes or generates additional hazards to a domain that is traditionally recognized as one of the most dangerous work environments.

Regulations: Worldwide aviation regulatory authorities have been constantly updating their policies and regulations with the aim of improving drone integration into their national airspace systems and expanding this technology's commercial adoption. For example, the Federal Aviation Administration (FAA) regulations in the US require drones to be operated at a maximum speed of 100 mph, a maximum height of 400 ft. above ground level (AGL) and a maximum horizontal distance of 400 ft. away from a structure [85]. In addition, the aerial vehicle must not exceed a weight of 55 lbs. including the payloads and should be only operated during daylight time, within the pilot-in-command's visual line of sight (VLOS), and over people who are directly involved in the drone flight [85]. Until very recently, a drone pilot had to go through a waiving process that, if approved, would enable drone operators to overcome some of these limitations. Recently, other countries have also started establishing rules pertaining to drone operation. For example, the European Union Aviation Safety Agency (EASA) recently developed a framework that categorizes drone operations into three – open, specific, and certified – depending on the risks involved with the flight operations [86]. The regulations mainly require drone operation to be performed by certified and registered pilots (except when operating camera-free drones weighing 250 g or less), within the VLOS, and at a maximum height of 400 ft. [86]. Flights are prohibited at night, over groups of people, nearby airports, objects, and in locations where privacy could be invaded [86]. Despite the efforts made by worldwide aviation authorities to regulate drone-mediated operations and safely integrate them into their national airspaces, construction emerged as a unique industry that requires further and more specific collaborative regulatory frameworks. For example, it is not uncommon for jobsite personnel and flight team members to consider all construction workers as part of the drone flight team to circumvent the FAA regulations that restrict drone operation over people and active areas. The deployment of drones is also associated with ethics- and privacy-related concerns within and outside the jobsite boundaries, particularly in urban environments and crowded cities [5, 14, 83]. On the jobsite, construction workers might exhibit disruptive behavior due to continuous drone monitoring, despite prior notification by management about the intended flight operations. Civil liberty intrusions might also stem from nearby individuals who did not agree to have their movements tracked, but rather expected to have some degree of privacy.

Collaborative partnerships need to be established to get specific flight legislation passed to operate aerial platforms in the dynamic and complex environments of the jobsites. Additional regulations must also target drone operation in indoor construction environments. This is particularly important, especially since certain construction tasks (e.g., structural inspections, jobsite safety monitoring) necessitate drone navigation in indoor and outdoor environments to provide a comprehensive and detailed site overview [5, 17]. It should be also mentioned that some of the newly emerged modalities described previously also require collocated human-drone types of interaction, justifying the need for specific flight legislations in this context. In addition, as of yet, no specific safety and health regulations have been established by regulatory agencies such as the Occupational Safety and Health Administration (OSHA) pertaining to the onsite drone usage, a factor that warrants additional exploration. Therefore, a program pre-evaluation followed by implementation and periodic assessments is required to pass adequate legislations that ensure worker safety while not compromising the drones' potential in construction. Once these legislations become well-established and recognized, the safety-, ethics-, and privacy-related risks associated with the onsite usage of this technology will be better categorized and regulated. Not only would this expand the technology's integration and effectiveness in the field, but it would also enable insurance companies to comprehensively identify these risks and incorporate them into new all-inclusive insurance policies. Such policies would ultimately guide construction companies in case of any potential drone-related liability.

Environment: Conditions such as inclement weather (e.g., windy or stormy environments), unexpected natural events, increased sunlight reflectivity, and shadowing effect add more complexity to the onsite drone operation and control [1, 4]. Aerial platform usage also depends on the flight space or jobsite conditions, which may be challenging, especially in crowded (e.g., a congested area characterized by an increased number of nearby buildings) and obstacle-dense environments (e.g., onsite presence of cranes and overhanging members) [4, 5]. For example, operating aerial platforms in such situations might: (1) cause drones to potentially lose their intended trajectory and come into contact with nearby jobsite objects, equipment, or machinery; (2) generate platform vibrations, affecting the resulting acquired visual quality; (3) affect the team members' visibility and control ability; and (4) deplete the platforms' battery capacities especially while stabilizing the drones in the air, significantly reducing the flight mission durations.

In light of these considerations, future research must be tailored toward improving the drones' performance in different weather and flight space conditions. More specifically, studies should focus on developing impact energy reduction and fail-safe systems to mitigate the hazards associated with potential drone failures in unanticipated and safety-critical conditions while ensuring a safe deployment [5]. Such systems range from autonomous navigation functions (e.g., auto-return home, auto-takeoff, and landing), sensor systems (e.g., obstacle avoidance, navigation system redundancy), and failsafe algorithms, to hardware components (e.g., protective cages, propeller guards) and recovery systems (e.g., parachute recovery system, airbags) [5, 87]. Although not designed to prevent drones from failing – as accidents are expected to happen – these systems are intended to minimize the safety risks associated with drone failure on people, environment, and other equipment.

6.6 Conclusion

Drones are expected to populate future construction jobsites. However, the success of such integration depends on several factors, which have been thoroughly explored and summarized in this chapter. Ideally, this chapter helps researchers and practitioners gain a comprehensive understanding of drones' current and future potential applications in construction and their advantages relative to conventional methods. It also summarizes the interaction roles that humans and drones could play during flight missions, the HDI research areas currently being explored, as well as this technology's implementation challenges. The chapter finally provides researchers and practitioners with a research roadmap to overcome these encountered barriers and guide them toward potential areas that warrant further research.

The literature reviewed showed that drones are being used throughout the entire construction lifecycle, with applications including site evaluation, site planning, mapping and earthwork volumetrics, construction progress monitoring, construction safety monitoring and inspection, structure and infrastructure inspection, post-disaster assessment, and marketing. In the near future, construction jobsite personnel will also rely on drones for security surveillance, aerial construction and material handling, as well as site communication types of tasks. Stated advantages include improvements in terms of time, cost, and safety compared to traditional procedures.

The literature also indicated that humans could play different active or passive roles when interacting with drones, depending on their level of involvement and the construction application context. These human roles include operators, supervisors, and recipients. On the other hand, drones could also act as social companions, personal assistants, and recipients. In terms of HDI research areas currently being explored, the literature revealed that researchers and practitioners investigate human-drone interaction modalities (i.e., gesture, speech, gaze, brain control, touch), comfortable approach distance, and trust. The factors potentially affecting trust between humans and drones were identified and consisted of improving humans' and drones' abilities to perceive, acknowledge, and naturally interact with each other; ensuring well-established communication between the interactants; and designing anthropomorphic and human-friendly types of drones.

Finally, the commonly encountered drone implementation challenges that were identified from the HDI and construction literature paved the way to developing a research roadmap mainly targeting four areas that warrant future exploration: drone (i.e., ground and onboard technologies), human (i.e., training, psychological effects), regulations (i.e., flight legislation, ethics and privacy), and environment (i.e., weather and flight space conditions).

References

1. G. Albeaino, M. Gheisari, B.W. Franz, A systematic review of unmanned aerial vehicle application areas and technologies in the AEC domain. J. Inf. Technol. Constr. ITcon. **24**, 381–405 (2019). https://www.itcon.org/2019/20
2. Z. Zhou, J. Irizarry, Y. Lu, A multidimensional framework for unmanned aerial system applications in construction project management. J. Manag. Eng. **34**, 04018004–04018004 (2018). https://doi.org/10.1061/(ASCE)ME.1943-5479.0000597
3. S. Zhou, M. Gheisari, Unmanned aerial system applications in construction: a systematic review. Constr. Innov., CI-02-2018–0010 (2018). https://doi.org/10.1108/CI-02-2018-0010

4. J.G. Martinez, M. Gheisari, L.F. Alarcón, UAV integration in current construction safety planning and monitoring processes: case study of a high-rise building construction project in Chile. J. Manag. Eng. **36** (2020). https://doi.org/10.1061/(ASCE)ME.1943-5479.0000761
5. J.G. Martinez, G. Albeaino, M. Gheisari, R.R.A. Issa, L.F. Alarcón, iSafeUAS: An unmanned aerial system for construction safety inspection. Autom. Constr. **125**, 103595 (2021). https://doi.org/10.1016/j.autcon.2021.103595
6. A. Ibrahim, M. Golparvar-Fard, 4D BIM based optimal flight planning for construction monitoring applications using camera-equipped UAVs, in *Computing in Civil Engineering 2019, Atlanta, Georgia*, (2019), pp. 217–224. https://doi.org/10.1061/9780784482438.028
7. I. Mutis, A.F. Romero, Thermal performance assessment of curtain walls of fully operational buildings using infrared thermography and unmanned aerial vehicles, in *Advances in Informatics and Computing in Civil and Construction Engineering*, ed. by I. Mutis, T. Hartmann, (Springer International Publishing, Cham, 2019), pp. 703–709
8. D. Roca, S. Lagüela, L. Díaz-Vilariño, J. Armesto, P. Arias, Low-cost aerial unit for outdoor inspection of building façades. Autom. Constr. **36**, 128–135 (2013). https://doi.org/10.1016/j.autcon.2013.08.020
9. M. Obaid, O. Mubin, C.A. Basedow, A.A. Ünlüer, M.J. Bergström, M. Fjeld, A drone agent to support a clean environment, in *Proceedings of the 3rd International Conference on Human-Agent Interaction*, (Association for Computing Machinery, New York, 2015), pp. 55–61. https://doi.org/10.1145/2814940.2814947
10. S. Schneegass, F. Alt, J. Scheible, A. Schmidt, Midair displays: concept and first experiences with free-floating pervasive displays, in *Proceedings of the International Symposium on Pervasive Displays*, (Association for Computing Machinery, New York, 2014), pp. 27–31. https://doi.org/10.1145/2611009.2611013
11. M. Avila, M. Funk, N. Henze, DroneNavigator: using drones for navigating visually impaired persons, in *Proceedings of the 17th International ACM SIGACCESS Conference on Computers & Accessibility*, (Association for Computing Machinery, New York, 2015), pp. 327–328. https://doi.org/10.1145/2700648.2811362
12. M. Al Zayer, S. Tregillus, J. Bhandari, D. Feil-Seifer, E. Folmer, Exploring the use of a drone to guide blind runners, in *Proceedings of the 18th International ACM SIGACCESS Conference on Computers & Accessibility*, (Association for Computing Machinery, New York, 2016), pp. 263–264. https://doi.org/10.1145/2982142.2982204
13. F. Mueller, M. Muirhead, Understanding the design of a flying jogging companion, in *Proceedings of the Adjunct Publication of the 27th Annual ACM Symposium on User Interface Software and Technology*, (Association for Computing Machinery, New York, 2014), pp. 81–82. https://doi.org/10.1145/2658779.2658786
14. M. Gheisari, B. Esmaeili, Applications and requirements of unmanned aerial systems (UASs) for construction safety. Saf. Sci. **118**, 230–240 (2019). https://doi.org/10.1016/j.ssci.2019.05.015
15. H. Golizadeh, M.R. Hosseini, D.J. Edwards, S. Abrishami, N. Taghavi, S. Banihashemi, Barriers to adoption of RPAs on construction projects: a task–technology fit perspective. Constr. Innov. **19**, 149–169 (2019). https://doi.org/10.1108/CI-09-2018-0074
16. J.G. Martinez, G. Albeaino, M. Gheisari, W. Volkmann, L.F. Alarcón, UAS point cloud accuracy assessment using structure from motion–based photogrammetry and PPK georeferencing technique for building surveying applications. J. Comput. Civ. Eng. **35**, 05020004 (2021). https://doi.org/10.1061/(ASCE)CP.1943-5487.0000936
17. R. Eiris, G. Albeaino, M. Gheisari, B. Benda, R. Faris, Indrone: visualizing drone flight patterns for indoor building inspection tasks, in *Proceedings of the 20th International Conference on Construction Applications of Virtual Reality*, (Teesside University Press, Middlesbrough, 2020), pp. 273–282. https://www.researchgate.net/publication/344444905_Enabling_the_Development_and_Implementation_of_Digital_Twins_Proceedings_of_the_20th_International_Conference_on_Construction_Applications_of_Virtual_Reality

18. C.H. Hugenholtz, J. Walker, O. Brown, S. Myshak, Earthwork Volumetrics with an Unmanned Aerial Vehicle and Softcopy Photogrammetry. J. Surv. Eng. **141**, 06014003 (2015). https://doi.org/10.1061/(ASCE)SU.1943-5428.0000138
19. P. Sestras, S. Roşca, Ş. Bilaşco, S. Naş, S.M. Buru, L. Kovacs, V. Spalević, A.F. Sestras, Feasibility assessments using unmanned aerial vehicle technology in heritage buildings: rehabilitation-restoration, spatial analysis and tourism potential analysis. Sensors **20**, 2054 (2020). https://doi.org/10.3390/s20072054
20. M.-C. Wen, S.-C. Kang, Augmented reality and unmanned aerial vehicle assist in construction management, in *Computing in Civil and Building Engineering 2014*, (American Society of Civil Engineers, Orlando, 2014), pp. 1570–1577. https://doi.org/10.1061/9780784413616.195
21. W. Jiang, Y. Zhou, L. Ding, C. Zhou, X. Ning, UAV-based 3D reconstruction for hoist site mapping and layout planning in petrochemical construction. Autom. Constr. **113**, 103137 (2020). https://doi.org/10.1016/j.autcon.2020.103137
22. S. Siebert, J. Teizer, Mobile 3D mapping for surveying earthwork projects using an Unmanned Aerial Vehicle (UAV) system. Autom. Constr. **41**, 1–14 (2014). https://doi.org/10.1016/j.autcon.2014.01.004
23. C.A.F. Ezequiel, M. Cua, N.C. Libatique, G.L. Tangonan, R. Alampay, R.T. Labuguen, C.M. Favila, J.L.E. Honrado, V. Canos, C. Devaney, A.B. Loreto, J. Bacusmo, B. Palma, UAV aerial imaging applications for post-disaster assessment, environmental management and infrastructure development, in *2014 International Conference on Unmanned Aircraft Systems (ICUAS)*, (IEEE, 2014), pp. 274–283. https://doi.org/10.1109/ICUAS.2014.6842266
24. J. Unger, M. Reich, C. Heipke, UAV-based photogrammetry: monitoring of a building zone. ISPRS – Int. Arch. Photogramm. Remote Sens. Spat. Inf. Sci. **XL–5**, 601–606 (2014). https://doi.org/10.5194/isprsarchives-XL-5-601-2014
25. U.S. Bureau of Labor Statistics, National Census of Fatal Occupational Injuries in 2017, 9 (2018)
26. M. Gheisari, A. Rashidi, B. Esmaeili, Using unmanned aerial systems for automated fall hazard monitoring, in *Construction Research Congress 2018*, vol. 2018, (American Society of Civil Engineers, New Orleans), pp. 62–72. https://doi.org/10.1061/9780784481288.007
27. National Equipment Register, Annual Theft Report (2016), 2017. http://www.ner.net/wp--content/uploads/2017/10/Annual-Theft-Report-2016.pdf
28. A. Bürkle, F. Segor, M. Kollmann, Towards autonomous micro UAV swarms. J. Intell. Robot. Syst. **61**, 339–353 (2011). https://doi.org/10.1007/s10846-010-9492-x
29. N. Michael, J. Fink, V. Kumar, Cooperative manipulation and transportation with aerial robots. Auton. Robots. **30**, 73–86 (2011). https://doi.org/10.1007/s10514-010-9205-0
30. Q. Lindsey, D. Mellinger, V. Kumar, Construction of cubic structures with quadrotor teams, in *Proceedings of the Robotics Science and Systems, Los Angeles, CA, USA*, (2011) https://doi.org/10.15607/RSS.2011.VII.025
31. S. Goessens, C. Mueller, P. Latteur, Feasibility study for drone-based masonry construction of real-scale structures. Autom. Constr. **94**, 458–480 (2018). https://doi.org/10.1016/j.autcon.2018.06.015
32. J. Irizarry, M. Gheisari, B.N. Walker, Usability assessment of drone technology as safety inspection tools. J. Inf. Technol. Constr. ITcon. **17**, 194–212 (2012)
33. C.-H. Yang, M.-C. Wen, Y.-C. Chen, S.-C. Kang, An optimized unmanned aerial system for bridge inspection, in *Proceedings of the 32nd International Symposium on Automation and Robotics in Construction and Mining ISARC 2015*, ed. by M. Malaska, R. Heikkilä, (International Association for Automation and Robotics in Construction (IAARC), Oulu, 2015), pp. 1–6. https://doi.org/10.22260/ISARC2015/0084
34. L. Duque, J. Seo, J. Wacker, Bridge deterioration quantification protocol using UAV. J. Bridge Eng. **23**, 04018080 (2018). https://doi.org/10.1061/(ASCE)BE.1943-5592.0001289
35. A.P. Tomiczek, T.J. Whitley, J.A. Bridge, P.G. Ifju, Bridge inspections with small unmanned aircraft systems: case studies. J. Bridge Eng. **24**, 05019003 (2019). https://doi.org/10.1061/(ASCE)BE.1943-5592.0001376

36. D. Kang, Y.-J. Cha, Autonomous UAVs for structural health monitoring using deep learning and an ultrasonic beacon system with geo-tagging. Comput.-Aided Civ. Infrastruct. Eng. **33**, 885–902 (2018). https://doi.org/10.1111/mice.12375
37. S.M. Adams, M.L. Levitan, C.J. Friedland, High resolution imagery collection for post-disaster studies utilizing unmanned aircraft systems (UAS). Photogramm. Eng. Remote Sens. **80**, 1161–1168 (2014). https://doi.org/10.14358/PERS.80.12.1161
38. N. Michael, S. Shen, K. Mohta, Y. Mulgaonkar, V. Kumar, K. Nagatani, Y. Okada, S. Kiribayashi, K. Otake, K. Yoshida, K. Ohno, E. Takeuchi, S. Tadokoro, Collaborative mapping of an earthquake-damaged building via ground and aerial robots. J. Field Robot. **29**, 832–841 (2012). https://doi.org/10.1002/rob.21436
39. G.-J.M. Kruijff, F. Pirri, M. Gianni, P. Papadakis, M. Pizzoli, A. Sinha, V. Tretyakov, T. Linder, E. Pianese, S. Corrao, F. Priori, S. Febrini, S. Angeletti, Rescue robots at earthquake-hit Mirandola, Italy: A field report, in *2012 IEEE International Symposium on Safety, Security, and Rescue Robotics (SSRR)*, (IEEE, 2012), pp. 1–8. https://doi.org/10.1109/SSRR.2012.6523866
40. D. Feil-Seifer, M.J. Matarić, Human-robot interaction, in *Encyclopedia of Complexity and Systems Science*, ed. by R. A. Meyers, (Springer, New York, 2009), pp. 4643–4659. https://doi.org/10.1007/978-0-387-30440-3_274
41. S. Mirri, C. Prandi, P. Salomoni, Human-drone interaction: state of the art, open issues and challenges, in *Proceedings of the ACM SIGCOMM 2019 Workshop on Mobile AirGround Edge Computing, Systems, Networks, and Applications*, (Association for Computing Machinery, New York, 2019), pp. 43–48. https://doi.org/10.1145/3341568.3342111
42. A. Yeh, P. Ratsamee, K. Kiyokawa, Y. Uranishi, T. Mashita, H. Takemura, M. Fjeld, M. Obaid, Exploring proxemics for human-drone interaction, in *Proceedings of the 5th International Conference on Human Agent Interaction*, (Association for Computing Machinery, New York, 2017), pp. 81–88. https://doi.org/10.1145/3125739.3125773
43. A. Wojciechowska, J. Frey, S. Sass, R. Shafir, J.R. Cauchard, Collocated human-drone interaction: methodology and approach strategy, in *2019 14th ACM/IEEE International Conference on Human-Robot Interaction (HRI)*, (2019), pp. 172–181. https://doi.org/10.1109/HRI.2019.8673127
44. D. Tezza, M. Andujar, The state-of-the-art of human–drone interaction: a survey. IEEE Access. **7**, 167438–167454 (2019). https://doi.org/10.1109/ACCESS.2019.2953900
45. C.A. Thiels, J.M. Aho, S.P. Zietlow, D.H. Jenkins, Use of unmanned aerial vehicles for medical product transport. Air Med. J. **34**, 104–108 (2015). https://doi.org/10.1016/j.amj.2014.10.011
46. K. Higuchi, T. Shimada, J. Rekimoto, Flying sports assistant: external visual imagery representation for sports training, in *Proceedings of the 2nd Augmented Human International Conference*, (Association for Computing Machinery, New York, 2011). https://doi.org/10.1145/1959826.1959833
47. E. Graether, F. Mueller, Joggobot: a flying robot as jogging companion, in *CHI '12 Extended Abstracts on Human Factors in Computing Systems*, (Association for Computing Machinery, New York, 2012), pp. 1063–1066. https://doi.org/10.1145/2212776.2212386
48. J.R. Cauchard, J.L. E, K.Y. Zhai, J.A. Landay, Drone & Me: An Exploration into Natural Human-Drone Interaction, in *Proceedings of the 2015 ACM International Joint Conference on Pervasive and Ubiquitous Computing*, (Association for Computing Machinery, New York, 2015), pp. 361–365. https://doi.org/10.1145/2750858.2805823
49. K. Miyoshi, R. Konomura, K. Hori, Above your hand: direct and natural interaction with aerial robot, in *ACM SIGGRAPH 2014 Emerging Technologies*, (Association for Computing Machinery, New York, 2014). https://doi.org/10.1145/2614066.2614086
50. J.R. Cauchard, A. Tamkin, C.Y. Wang, L. Vink, M. Park, T. Fang, J.A. Landay, Drone.io: a gestural and visual interface for human-drone interaction, in *Proceedings of the 14th ACM/IEEE International Conference on Human-Robot Interaction*, (2019), pp. 153–162. https://doi.org/10.1109/HRI.2019.8673011

51. A. Mashood, H. Noura, I. Jawhar, N. Mohamed, A gesture based kinect for quadrotor control, in *2015 International Conference on Information and Communication Technology Research (ICTRC)*, (2015), pp. 298–301. https://doi.org/10.1109/ICTRC.2015.7156481
52. A. Sanna, F. Lamberti, G. Paravati, F. Manuri, A Kinect-based natural interface for quadrotor control. Entertain. Comput. **4**, 179–186 (2013). https://doi.org/10.1016/j.entcom.2013.01.001
53. J.L. E, I.L. E, J.A. Landay, J.R. Cauchard, Drone & Wo: cultural influences on human-drone interaction techniques, in *Proceedings of the 2017 CHI Conference on Human Factors in Computing Systems*, (Association for Computing Machinery, New York, 2017), pp. 6794–6799. https://doi.org/10.1145/3025453.3025755
54. R.A.S. Fernández, J.L. Sanchez-Lopez, C. Sampedro, H. Bavle, M. Molina, P. Campoy, Natural user interfaces for human-drone multi-modal interaction, in *2016 International Conference on Unmanned Aircraft Systems (ICUAS)*, (2016), pp. 1013–1022. https://doi.org/10.1109/ICUAS.2016.7502665
55. M. Quigley, M.A. Goodrich, R.W. Beard, Semi-autonomous human-UAV interfaces for fixed-wing mini-UAVs, in *2004 IEEE/RSJ International Conference on Intelligent Robots and Systems (IROS) (IEEE Cat. No.04CH37566)*, vol. 3, (2004), pp. 2457–2462. https://doi.org/10.1109/IROS.2004.1389777
56. G. Jones, N. Berthouze, R. Bielski, S. Julier, Towards a situated, multimodal interface for multiple UAV control, in *2010 IEEE International Conference on Robotics and Automation*, (2010), pp. 1739–1744. https://doi.org/10.1109/ROBOT.2010.5509960
57. B.N. Pavan Kumar, A. Balasubramanyam, A.K. Patil, B. Chethana, Y.H. Chai, GazeGuide: an eye-gaze-guided active immersive UAV camera. Appl. Sci. **10**, 1668 (2020). https://doi.org/10.3390/app10051668
58. M. Staudte, M.W. Crocker, Investigating joint attention mechanisms through spoken human–robot interaction. Cognition **120**, 268–291 (2011). https://doi.org/10.1016/j.cognition.2011.05.005
59. J.P. Hansen, A. Alapetite, I.S. MacKenzie, E. Møllenbach, The use of gaze to control drones, in *Proceedings of the Symposium on Eye Tracking Research and Applications*, (Association for Computing Machinery, New York, 2014), pp. 27–34. https://doi.org/10.1145/2578153.2578156
60. L. Yuan, C. Reardon, G. Warnell, G. Loianno, Human gaze-driven spatial tasking of an autonomous MAV. IEEE Robot. Autom. Lett. **4**, 1343–1350 (2019). https://doi.org/10.1109/LRA.2019.2895419
61. N. Kosmyna, F. Tarpin-Bernard, B. Rivet, Towards brain computer interfaces for recreational activities: piloting a drone, in *Human-Computer Interaction – INTERACT 2015*, ed. by J. Abascal, S. Barbosa, M. Fetter, T. Gross, P. Palanque, M. Winckler, (Springer International Publishing, Cham, 2015), pp. 506–522
62. B.H. Kim, M. Kim, S. Jo, Quadcopter flight control using a low-cost hybrid interface with EEG-based classification and eye tracking. Comput. Biol. Med. **51**, 82–92 (2014). https://doi.org/10.1016/j.compbiomed.2014.04.020
63. K. LaFleur, K. Cassady, A. Doud, K. Shades, E. Rogin, B. He, Quadcopter control in three-dimensional space using a noninvasive motor imagery-based brain-computer interface. J. Neural Eng. **10**, 046003–046003 (2013). https://doi.org/10.1088/1741-2560/10/4/046003
64. c.W. Mueller, c. D'Andrea, Critical subsystem failure mitigation in an indoor UAV testbed, in *2012 IEEE/RSJ International Conference on Intelligent Robots and Systems*, (2012), pp. 780–785. https://doi.org/10.1109/IROS.2012.6385910
65. P. Abtahi, B. Landry, J. (Junrui) Yang, M. Pavone, S. Follmer, J.A. Landay, Beyond the force: using quadcopters to appropriate objects and the environment for haptics in virtual reality, in *Proceedings of the 2019 CHI Conference on Human Factors in Computing Systems*, (Association for Computing Machinery, New York, 2019), pp. 1–13. https://doi.org/10.1145/3290605.3300589
66. P. Abtahi, D.Y. Zhao, J.L. E, J.A. Landay, Drone near me: exploring touch-based human-drone interaction, in *Proceedings of the ACM on Interactive, Mobile, Wearable and Ubiquitous Technologies*, vol. 1, (2017). https://doi.org/10.1145/3130899

67. K. Nitta, K. Higuchi, J. Rekimoto, HoverBall: augmented sports with a flying ball, in *Proceedings of the 5th Augmented Human International Conference*, (Association for Computing Machinery, New York, 2014). https://doi.org/10.1145/2582051.2582064
68. E.T. Hall, Distances in man, in *Hidden dimens*, (Anchor Books, New York, 1969)
69. B.A. Duncan, R.R. Murphy, A preliminary model for comfortable approach distance based on environmental conditions and personal factors, in *2012 International Conference on Collaboration Technologies and Systems (CTS)*, (2012), pp. 622–627. https://doi.org/10.1109/CTS.2012.6261117
70. U. Acharya, A. Bevins, B.A. Duncan, Investigation of human-robot comfort with a small Unmanned Aerial Vehicle compared to a ground robot, in *2017 IEEE/RSJ International Conference on Intelligent Robots and Systems (IROS)*, (2017), pp. 2758–2765. https://doi.org/10.1109/IROS.2017.8206104
71. W. Jensen, S. Hansen, H. Knoche, Knowing you, seeing me: investigating user preferences in drone-human acknowledgement, in *Proceedings of the 2018 CHI Conference on Human Factors in Computing Systems*, (Association for Computing Machinery, New York, 2018), pp. 1–12. https://doi.org/10.1145/3173574.3173939
72. W.S. Ng, E. Sharlin, Collocated interaction with flying robots, in *2011 RO-MAN*, (2011), pp. 143–149. https://doi.org/10.1109/ROMAN.2011.6005280
73. J.R. Cauchard, K.Y. Zhai, M. Spadafora, J.A. Landay, Emotion encoding in human-drone interaction, in *2016 11th ACM/IEEE International Conference on Human-Robot Interaction (HRI)*, (2016), pp. 263–270. https://doi.org/10.1109/HRI.2016.7451761
74. G. Morgenthal, N. Hallermann, Quality assessment of Unmanned Aerial Vehicle (UAV) based visual inspection of structures. Adv. Struct. Eng. **17**, 289–302 (2014). https://doi.org/10.1260/1369-4332.17.3.289
75. Y. Ham, K.K. Han, J.J. Lin, M. Golparvar-Fard, Visual monitoring of civil infrastructure systems via camera-equipped Unmanned Aerial Vehicles (UAVs): a review of related works. Vis. Eng. **4**, 1 (2016). https://doi.org/10.1186/s40327-015-0029-z
76. S. Kim, J. Irizarry, D.B. Costa, Field test-based UAS operational procedures and considerations for construction safety management: a qualitative exploratory study. Int. J. Civ. Eng. **9** (2020). https://doi.org/10.1007/s40999-020-00512-9
77. S. Alizadehsalehi, I. Yitmen, T. Celik, D. Arditi, The effectiveness of an integrated BIM/UAV model in managing safety on construction sites. Int. J. Occup. Saf. Ergon. **0**, 1–16 (2018). https://doi.org/10.1080/10803548.2018.1504487
78. S. Kim, J. Irizarry, D.B. Costa, Potential factors influencing the performance of Unmanned Aerial System (UAS) integrated safety control for construction worksites, in *Construction Research Congress 2016*, (American Society of Civil Engineers, San Juan, 2016), pp. 2614–2623. https://doi.org/10.1061/9780784479827.260
79. C. Eschmann, C.-M. Kuo, C.-H. Kuo, C. Boller, Unmanned aircraft systems for remote building inspection and monitoring, in *6th European Workshop on Structural Health Monitoring – Th2B1*, (2012) http://www.ndt.net/?id=14139
80. C. Mendes, D. Costa, R. Melo, Evaluating Uas-Image Pattern Recognition System Application For Safety Guardrails Inspection, 1–3 (2018)
81. R.R.S. de Melo, D.B. Costa, J.S. Álvares, J. Irizarry, Applicability of unmanned aerial system (UAS) for safety inspection on construction sites. Saf. Sci. **98**, 174–185 (2017). https://doi.org/10.1016/j.ssci.2017.06.008
82. R.R.S. de Melo, D.B. Costa, Integrating resilience engineering and UAS technology into construction safety planning and control. Eng. Constr. Archit. Manag. **26**, 2705–2722 (2019). https://doi.org/10.1108/ECAM-12-2018-0541
83. M. Herrmann, Unmanned aerial vehicles in construction: an overview of current and proposed rules. Constr. Res. Congr. **2016**, 588–596 (2016). https://doi.org/10.1061/9780784479827.060
84. J. Irizarry, D.B. Costa, Exploratory study of potential applications of unmanned aerial systems for construction management tasks. J. Manag. Eng. **32**, 05016001 (2016). https://doi.org/10.1061/(ASCE)ME.1943-5479.0000422

85. US Department of Transportation, Advisory Circular 107-2 – Small Unmanned Aircraft Systems (sUAS), 2016. https://www.faa.gov/documentlibrary/media/advisory_circular/ac_107-2.pdf.
86. European Union Aviation Safety Agency, Drones Information Notices 2020. Eur. Union Aviat. Saf. Agency. (2020). https://www.easa.europa.eu/document-library/general-publications/drones-information-notices-2020. Accessed 11 Feb 2021
87. G. Albeaino, M. Gheisari, Trends, benefits, and barriers of unmanned aerial systems in the construction industry: a survey study in the United States. J. Inf. Technol. Constr. ITcon. **26**, 84-111 (2021). https://doi.org/10.36680/j.itcon.2021.006

Chapter 7
Safety Challenges of Human-Drone Interactions on Construction Jobsites

Idris Jeelani and Masoud Gheisari

Abstract Over the last decade, there is an ever-increasing demand for automation in construction to counter the major challenges of low productivity and lack of skilled labor that have marred the construction industry for years. As a result, modern construction jobsites will feature several aerial robotic systems working side by side with human workers. The use of aerial robots (a.k.a. drones) or unmanned areal vehicles) in construction has already increased significantly, making the construction industry one of the fastest commercial adopters of drones. These aerial robots can be adopted through different phases of a construction project, from site surveying and mapping and progress monitoring to building inspection and structure maintenance. With the predicted increase in construction activity to address growing infrastructural needs, it is expected that collaborative interactions between human workers and drones in the construction industry will continue to increase rapidly. Since construction is still a worker-driven industry, drones will have to work symbiotically with human workers. Fully autonomous drones will coordinate their behaviors with human team members; they will not only be used as mere tools but rather will engage in a two-way interaction with workers to become quasi team members in the worker-drone teams. This collaboration and continuous interactions between human workers and drones raises novel occupational safety and health issues for construction workers, which are critical to identify, understand, and evaluate. Construction is already one of the most dangerous industries, and understanding the safety challenges of human interactions with drones in drone-populated construction sites will help with the successful integration of these technologies while minimizing their risks to the health and safety of construction workers. In this chapter, we first envision the drone-populated jobsites of the future and discuss the interactions in human-drone teams on the construction jobsites. We then discuss the novel health and safety risks that result from human interaction with drones. We mainly focus on physical risks, attentional allocation (e.g., visual, auditory, and

I. Jeelani (✉) · M. Gheisari
M.E. Sr Rinker School of Construction Management, University of Florida, Gainesville, FL, USA
e-mail: idris.jeelani@ufl.edu; masoud@ufl.edu

H. Jebelli et al. (eds.), *Automation and Robotics in the Architecture, Engineering, and Construction Industry*, https://doi.org/10.1007/978-3-030-77163-8_7

cognitive distractions), and psychological impacts (e.g., acute stress, cognitive and sensory overload, and negative emotional states) of such interactions on the human workers who work directly or indirectly with or around drones on the construction jobsites.

Keywords Human-drone interaction · Collaborative robots · Safety impact of drones

7.1 Introduction

Drones, a.k.a. unmanned aerial vehicles (UAVs), are becoming increasingly popular in several industries. In 2016, the global commercial drone market was worth \$5.8 billion; trends indicate that it will be worth more than \$130 billion in 2025 [1]. Originally intended for military use, drones have, in the last 10 years, evolved remarkably in terms of their hardware and software capabilities, which has made them increasingly valuable for civilian use. Drones are now commonplace in a number of industries, e.g., surveying, agriculture, and mining. Within the construction industry, the growth of drone use has been incredible. Drones have moved from what was essentially a novelty camera to one of the central tools of construction. Drone use in construction expanded by over 200% in a single year, which makes the construction industry the most rapid in terms of commercial adoption of drones [2].

Drones have proven to be incredibly useful to the construction industry due to their agility and customizability that allows construction managers and planners to reach different areas of construction sites much easier and faster than human beings and carry out their monitoring of inspection tasks using multiple sensors that drones can carry. Drone's small size and ability to fly in small spaces enable them to undertake task more swiftly and less expensively than other ground vehicles [3]. Additionally, new technologies enhancing the battery life, reliability of control, and GPS navigation have enabled the development of modern lightweight and low-cost aerial devices [4] that are increasingly being used in the industry [5–7]. Within the construction industry, drones are employed to help in project planning by conducting aerial mapping of construction sites, monitoring the construction workflow and site logistics and also to inspect sites for the assessment of the integrity of structures and to assess maintenance requirements. Drones are also employed in a number of other applications in the AEC (architecture, engineering, and construction) sector, which include traffic monitoring [8], monitoring landslides [9, 10], preserving cultural heritage [10, 11], and town planning [12, 13]. As drones become increasingly more autonomous, their influence in the way the construction activities are carried out will increase considerably. Drones are becoming an essential part of the construction toolkit, and in the very near future, it will be rare to see a major construction site without drones being an element of the workplace.

Even though there has been an outburst in the use of drones in the construction industry, it is still a human labor-driven industry. As construction continues to grow in line with national requirements for new infrastructure, there will clearly be greater interaction between human employees and autonomous actors like drones. Integrating these flying machines into the already risky workplace of construction workers inevitably creates new questions of health and safety for human workers. Even before the introduction of drones, construction was one of the nation's most perilous industries [14], with over 28,000 construction workers being killed in the workplace in the USA in the last 20 years [15]. In 2018 alone, the US Department of Labor states that 1038 construction workers died and more than 200,000 were injured, amounting to around 20% of deaths in the workplace in the industrial sector [16]. Construction workers already operate within a high-risk environment, and the introduction of drones may increase these risks.

As drones are becoming so rapidly ubiquitous, it is vital that the additional risks they bring to the workplace be evaluated. Gheisari and Esmaeili [17] undertook recent research in the USA looking at a number of logistical, economic, legal, and societal elements that could influence the deployment of drones in the construction industry. Two primary concerns that were noted were related to the safety issues in integrating drones in construction and also legal and liability questions. In terms of legality and liability, problems could arise related to insurance, property rights, the law of trespass, invasion of privacy, and any personal injuries or damage to properties as a result of errors in drone operation. In terms of safety, it is clear that deploying flying machinery on construction sites could create additional safety risks. These risks range from the possibility of drones colliding with each other or other machinery, colliding with workers, or causing workers to become distracted or stressed. Such potential dangers to health and safety require robust investigation and the development of proper safety protocols so that drones can operate safely alongside human workers and other construction machineries.

Although much research has been undertaken into how drones can be deployed in construction and their beneficial impact, there is a paucity regarding the health and safety aspects of such deployments. It is vital that detailed research should be undertaken for evaluation of the risks drones may pose to human workers that work alongside them or in their environment. Prior to any empirical research being undertaken to assess the specific safety problems drones may pose, it is crucial that we create a general understanding of the ways in which worker's health and safety may be impacted by drones.

7.2 Drones and Safety in Construction

7.2.1 *Safety Issues in Construction*

Construction in the USA has an annual spend of $1231 billion, making it a central driver of the country's economy [18]. More than 7.3 million workers are employed in the sector, which supplies 4.2% of gross domestic product (GDP) [19]. On the other hand, construction is also one of the most hazardous industries for workers [14]. The US Bureau of Labor Statistics, in its 2018 figures, reported that there were more than 1000 fatalities and 200,000 non-fatal injuries in the industry [16]. More than 30% of fatalities in the workplace in the USA are in construction, even though only 7% of the workforce are employed in the sector [20]. In spite of advances in technology and regulation, industry fatalities have held steady at around 900 per year [15, 16, 20, 21]. In addition to fatalities, non-fatal injuries in the sector have also kept a consistently high rate at more than 200,000 per year [16, 20, 21]. Alongside the personal and societal distress such as injuries caused to the workers themselves and their families, there is a very high financial price to pay in addition. It has been estimated that the combined costs of fatalities and injuries in the construction sector are in excess of $48 billion, significantly reducing revenue and profits for construction businesses and also causing reputational damage [22].

The disproportionately high rates of injury in the construction sector occur for a variety of reasons. In contrast to other industrial sectors, in which work is generally undertaken as a single location, such as a factory, construction frequently takes place in a workplace that is temporary, dynamic, and subject to rapid change [23]. The majority of industrial workers have specifically delineated tasks to undertake that they repeat within the same environment, whereas construction workers often have to cope with previously unpredicted issues and safety hazards [24]. This means that construction workers are exposed to a multiplicity of hazardous situations making it difficult to carry out generalized hazard assessments and implement fail-safe control processes. Lastly, construction sites generally involve a number of different contractors, so that responsibility for safety is decentralized, making it more likely that hazards will be disregarded [24]. All these specific elements increase the challenges of maintaining a safe environment in the construction industry, which results in disproportionate fatality/injury statistics as seen in Fig. 7.1. As a result, construction workers need to be more attentive and cognizant of their workplace environment and be more alert to the potential of unexpected dangers.

Although risk levels are high for every construction worker, analysis has shown that the greatest number of fatalities in the sector occurs among speciality contractors undertaking specific jobs such as excavation, structural steelwork, masonry/concrete work, electrical, painting, plumbing, and heating, ventilation, and air conditioning (HVAC) [16, 20, 21]. Workers in these specific trades frequently work within dangerous environments using heavy machinery and potentially hazardous tools; additionally, they are frequently working against the clock. Among others, one of the most common causes of fatalities in these areas has been falls, accounting

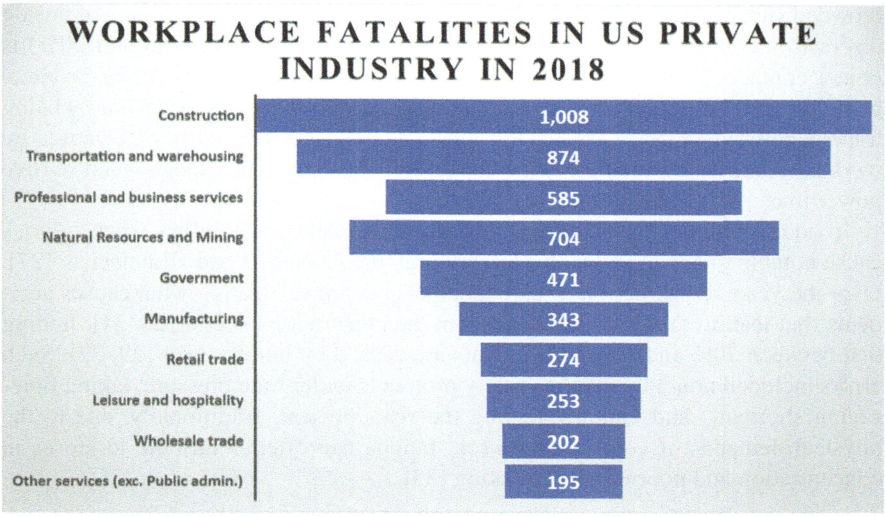

Fig. 7.1 Occupational fatalities 2018. (Data source: US Bureau of Labor Statistics)

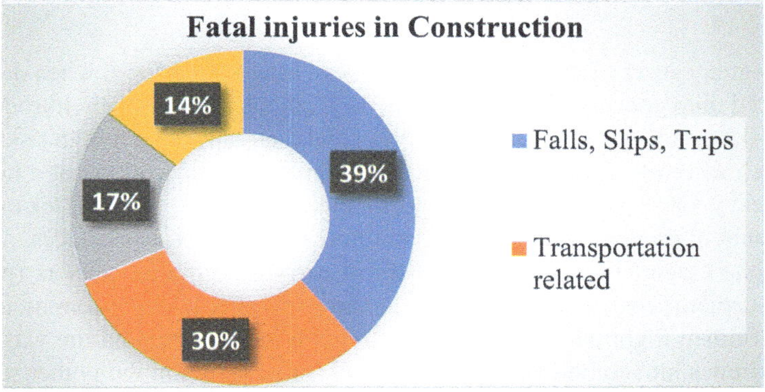

Fig. 7.2 Fatalities in construction by accident type. (Data source: US Bureau of Labor Statistics [15])

for approximately 30% of fatalities [25] (Fig. 7.2). This is because construction sites by their nature generally involve under construction structures characterized by uneven surfaces, inadequate fall protection, and temporary working platforms placed at a significant height from the ground. Statistically, construction workers are at the greatest risk of falling when working on ladders, the edges of roofs, and scaffolding [26]. Following these, the next greatest risk of fatality is accidents caused by transportation, accounting for approximately 25% of fatalities [25]. Construction sites generally involve the moving of various pieces of heavy machinery such as trucks, cranes, and excavators. These equipment often operate in crowded sites with other equipment and workers increasing the likelihood of unsafe physical interactions. The next most common cause of fatalities (around 20%) is object contact [25] (Fig. 7.2), with the most frequent occurrence being workers being hit with falling objects and then being trapped by cave-ins, followed by being trapped between objects [24]. Lastly, approximately 15% of fatalities are caused by workers coming into contact with dangerous substances or objects such as live power lines [25] (Fig. 7.2).

In addition to causing pain and distress to workers and families, such injuries cause notable financial losses due to very high medical costs and absenteeism [27]. Over the years, construction researchers have been investigating what causes accidents that lead to fatalities and injuries in the construction sector [28–31], finding that between 70% and 80% of accidents are caused by human error [29, 32]. Such errors included non-adherence to safety protocols under time pressure, taking time-saving shortcuts, and underestimating the risks present. Additionally, due to the physical demands of construction work, fatigue can often contribute to lapses in concentration and poor decision-making [33].

7.2.2 Drones in Contemporary Construction Environments and Potential Future Uses

The potential causes of accidents detailed above may be amplified when drones are introduced onto construction sites, where they can either potentially introduce new safety risks or act as a distraction to workers who already have a high risk of accident, such as those operating in a hazardous environment, working at a height, or using heavy or dangerous machinery. In this subsection, we will examine the deployment of drones within current and future construction workplaces.

A drone essentially is an aerial platform carrying different sensors (predominately a camera) controlled remotely by a human pilot or flying autonomously on a pre-determined flight path [34]. The construction industry has led the way in integrating drones into commercial operations [2]. In comparison with other sectors of AEC, construction has the highest and most rapidly expanding rate of drone deployment [35]. Drones are used for monitoring progress [36], creating maps and surveys [37], inspecting buildings [38], monitoring for safety issues [39], handling materials [40], communicating around sites [41], aerial construction [42], and site security purposes [43]. Two common types of commercial drones used in the industry are fixed-wing and rotary-wing drones with rotary-wing aircraft being used more frequently than fixed-wing ones [17]. It is predominantly because they are flexible, capable of vertical takeoff and landing, and capable of hovering and landing safely after an engine failure, which increases the popularity and efficiency in a number of areas of construction [4]. Table 7.1 lists the top identified technology needs and components based on a national survey.

Table 7.1 Top drone technology needs/components in the current construction work context, adapted from [3, 17]

Technology needs/ components	*Description*
Vehicle type: *Rotary-wing*	Composed of one or multiple propellers; do not require runways, as they can take off and land vertically; high hovering capabilities; considered safe and stable due to propellers redundancy; low flying endurance
Flying style: *Autonomous flights characteristics used*	Auto-return home; auto take-off/landing; obstacle avoidance; waypoint navigation; guarded motion; follow target
Onboard sensors	Visual cameras; thermal cameras; motion detection; LiDAR; audio sensors; RFID readers; air quality sensors

Drones already influencing infrastructure design, construction, operation and maintenance, and all indicators point toward the continued expansion of the use of drones within construction. Internet of Things (IoT), wireless technology, and new sensing devices will create "smart" construction processes, where drones play a vital role. Drones employing sensors, e.g., LiDAR, GPS, infrared and visual cameras, etc., will be able to capture crucial data from the construction site for effective monitoring and making decisions related to construction progress and quality. With artificial intelligence (AI)-based systems, drones can help construction workers in their work, for example, by guiding the operators of tower cranes, overseeing safety procedures, and carrying out quality control activities. Drones will inevitably be used more and more for these collaborations [3]. As construction remains highly dependent on human workers, drones will have to work symbiotically with human construction workers. As a result, the role of drones is changing from passive (observational roles) to active (interacting with humans). In the future, fully autonomous drones would engage and communicate with human workers as effective team members rather than mere tools of data collection. Drones will offer assistance to humans in several aspects of construction, such as watching blind spots when cranes are being operated and carrying materials up to workers at high levels in addition to performing several tasks autonomously such as inspection, safety monitoring, delivery, and building, collaborating with humans to undertake a number of common tasks (Table 7.2).

7.2.3 *Drone Safety Within Construction Work*

Although it is indisputable that the construction industry can derive great benefits from drone deployment, it is crucial to acknowledge the risks to health and safety that drones can impose and fully understand the human, societal, and economic costs that could arise from their use. Such risks arise from the unintended physical interaction between workings and drones or subsequent effects of being distractions

Table 7.2 Various types of drones and their potential uses

Drone types	Description	Uses
Builder drones	Drone connected to robotic arms and material reservoirs to perform different tasks	Performing construction tasks such as spray painting, bricklaying, aligning reinforcement bars, and laying pipes and conduits
Delivery drones	Robust-framed drones with high-performance rotors/engines for carrying heavier loads	Delivering tools and light materials (e.g., fasteners, coupling joints, nails, screws to workers at heights) Moving material, documents, etc. within the construction site
Safety drones	Drones carrying visual/infrared sensors, computers for live streaming data, with remote service processing this data employing AI and algorithms for the identification of potential safety hazards	Conducting aerial inspections using predefined flight paths, with real-time data processing taking place service and safety managers being alerted to any issues Safety drones will offer assistance to crane operators, safely helping them transport materials and monitoring any blind spots
Inspector drones	Drones carrying GPS, high-risk cameras, and other localized systems	Conducting an autonomous inspection of building works, streaming imagery to servers that will employ computerized analysis to assess progress and alert operators to any quality control problems

caused by flying drones. Every time humans and drones interact, whether cognitively or physically, there is a potential hazard. Additionally, multiple drones could be operating in the same airspace, resulting in workers facing potential hazards from a multiplicity of directions. As previously noted, the construction environment already poses far greater risks to workers than exists in other industrial sectors. Adding drones to such environments increases the extant risk and creates new ones. As discussed, a construction worker must have a high level of focus and attention as they work in unpredictable and high-risk environments. However, drones can affect workers' attention by causing unnecessary distractions which can have serious implications for the safety of workers and those working around them. Workers at heights who already have the highest fatality rates are particularly exposed to the risks imposed by drones. Another group of workers who have a high fatality rate are those using and sharing their work environment with heavy machinery, and the distracting potential of drones could also increase their risk.

A number of studies looking at safety issues around deploying drones on construction sites have identified several challenges that could impose limitations on the ways in which drones could be safely deployed in these environments (Table 7.3). These include the presence of multiple obstacles (cranes, trees, buildings), environmental challenges (bright sunlight, high winds, rain, snow, fog), technical

Table 7.3 Drone limitations with the potential to cause hazards

Limitation	References in the literature
Flight zone restrictions/pilot qualification	[44] [17] [45] [46] [47] [48] [49] [50]
Technical limitations (e.g., signal interference, low battery life, insufficient quality of visual sensors)	[51] [39] [17] [49] [45] [50] [47] [52] [53] [46]
Challenging environmental/weather conditions (e.g., snow, rain, wind, bright sun Wootton)	[39] [52] [50] [52] [47]
On-site obstacles (e.g., buildings, trees, cranes)	[39] [39] [46] [47]
Collisions with workers	[39] [17] [46] [47] [54] [55]
Workers becoming distracted	[39] [39] [17] [46] [47] [54] [55]

challenges (visual sensors of insufficient quality, low battery life, interference with signals), and regulatory challenges (qualifications for pilots, restricted flight zones).

Regrettably, no research has undertaken a comprehensive evaluation of the safety hazards caused by integrating drones into the construction industry, leaving significant gaps in the knowledge in this area. The lack of such research has resulted in an absence of drone-specific occupational health and safety regulations for construction sites. Drones are only minimally regulated by the Federal Aviation Administration (FAA), which only imposes general regulations on the commercial use of drones with no specific regulations for the complexities of construction sites. Furthermore, these regulations are only designed for the prevention of physical accident, with no consideration given to the way in which indirect accident-causing influences may be present, e.g., causing anxiety, stress, or distracting workers.

7.3 Drone Safety Challenges

As detailed in the previous discussion, we must acknowledge that drones present a number of risks for construction workers, either those who have direct interaction with them or those who work in the same environment. A drone is a physical entity on the construction site, and thus there is always a risk that it can collide with workers or other equipment or structure. This may occur if a drone or part of it strikes a worker or if technical issues result in drone itself or payload, falling on workers below its flight path. Drones can cause worker distraction, which can influence their capacity for safe working and thus increase the risk to themselves or their fellow workers. Lastly, a drone that works alongside humans, inspecting and recording their work, can have a negative psychological influence on these workers, triggering undesirable physiological reactions. These risks are discussed further in the sections that follow.

7.3.1 Physical Risks

A drone is a physical object that occupies and operates within the comparatively crowded airspace of any construction site, sharing the space with human operatives; structures, both permanent and temporary; and other equipment. Because of this, there is always a risk of drones coming into unintended physical contact with humans, buildings, or equipment resulting in accident or injury. Such injury could occur because of a number of sources, including:

- Engineering error: Faults within the hardware or software of the drone.
- Human error: Errors made by the flight team when piloting a drone or preparing and planning for a flight.
- Environmental stresses: Disturbances to safe flight as a result of weather and other environmental factors.
- Cyberattack: Nefarious actors, e.g., terrorists, may attempt to hack drones to deliberately cause injury or fatality.

An engineering error may encompass hardware errors (such as electronic faults, loose connections) and software errors (such as signal interference, faulty algorithms, errors in programming) [56]. As a result, a drone might begin operating unsafely, e.g., being unable to stop, deviating from the intended flight path, flying too fast without control, or making certain movements or accelerations. Human errors can encompass errors when directly piloting the drone or errors when planning and preparing for flight (such as not checking aircraft batteries, improper sense of connection, not accounting for objects that obstruct the planned flight path) [56]. This type of human error can cause drones to move in unexpected directions or to fall from the sky and come into unintended physical contact with workers. It should be noted that human errors most frequently result from operatives becoming overfamiliar with their equipment and placing too much trust in it [57]. Within the construction sector, this sort of overconfidence/trust issue could lead to workers positioning themselves in a manner that imperils their safety. With the construction environment frequently lacking stability and being dynamic and fluid, new scenarios are constantly arising, meaning that it is difficult to stick to previously trusted routines without increasing risk levels. In these scenarios, unanticipated movements by a drone can be responsible for significantly increasing the risk of injuries or even causing fatalities.

In addition to aforementioned human or engineering errors, poor weather or dangerous environmental conditions (inadequate lighting, high wind, heavy snow/rain, fog) may also disturb the intended flight path of a drone, resulting in unintentional physical divergence and possible injury/fatality [56]. Drone rotors, which spin at a high speed, can create gusts of wind that result in particulates being blown around from piles of gravel, sand, dirt, or loose soil, which are frequently present in the construction sites. The dust and particulates created in this way can be destructive to a drone's sensors and could cause it to malfunction. Such particulates and dust could, additionally, be directly damaging to health and safety for any nearby workers. Lastly, drone manufacturers and operators are becoming increasingly concerned that drones could be hacked for hostile purposes. Drones generally navigate via GPS and communicate with their controllers via Wi-Fi, both mechanisms carrying the risk of being hacked. Incidents have occurred where drones have lost control due to cyberattack by hostile actors, for example, through spoofing of GPS data [58].

Listed below are the various risks of physical contact posed by drones when they are deployed in the construction environment:

7.3.1.1 Struck by Accidents

Pilot error, interrupted connections, or environmental factors can all cause accidental collisions between drones and workers. Such collisions may cause physical injury to those directly struck by the drone and/or to those close by, and property may be damaged as well. When workers share their workspace with drones, it increases the likelihood of unintentional collisions resulting in injury. Workers who have direct interactions with drones face the greatest risk of injury as it is not possible for them to be isolated from the drone and still carry out their tasks. These workers include those working with drones in a team for the accomplishment of a variety of tasks, for example, delivering materials and building structures.

As drones become more prevalent in the construction environment, so the chance increases that they may crash into each other and into other equipment (e.g., cranes) and structures on-site or even suffer bird strike, all of which could result in the drone, parts of it, or its payload dropping onto workers below. This can also occur if there is a drone malfunction such as loss of signal or battery power. Workers below the drone flight path are particularly at risk of injury from such unanticipated hazards.

7.3.1.2 Caught-in or Caught-between Accidents

As previously noted, drones will often be required to work closely with human workers and share the same space, increasing hazard levels. Drones can now work within complex environments and undertake specialist tasks over and above their previous use of making simple video recordings for reconnaissance. These operations require high levels of autonomy and many moving parts, thus increasing the risk of workers having body parts trapped in the rotors or other mechanisms. Thus, in addition to the increased likelihood of struck-by incidents (i.e., drone colliding with a human worker), a close collaboration between humans and drones may also result in caught-in or caught-between accidents, where workers' body parts (such as hands or fingers) get caught in the fast-moving mechanical parts of drones (such as rotors).

7.3.1.3 Indirect/Secondary Accidents

Additionally, a drone may collide with other elements of the construction site, for example, electrical equipment or a live powerline, that could increase the risk of electrocution for any workers present. A number of incidents of this type have been noted in the literature [59, 60]. Similarly, drones can be in collision with unstable temporary structures, for example, scaffolds, which could cause them to fall, injuring or killing workers on such platforms or those working below.

7.3.1.4 Dust and Particulates

Drone rotors move extremely fast and can blow up significant amounts of dust when flying at a low level or close to stockpiles of bulk materials, as well as when taking off and landing. Such dust exposure can cause particle infiltration into human lungs, which can then become irritated, creating inflammatory reactions that ultimately impair oxygen diffusion and thus lung function [61]. Long-term dust exposure can cause severe chronic respiratory problems for those working on construction sites. Additionally, dust emissions can lead to a reduction in visibility, potentially compromising the ability of those operating equipment in the vicinity to control their machinery and thus increasing risks for workers.

7.3.2 Attentional Cost

Between 70% and 80% of accidents on construction sites are attributable to human error [29, 32]. Human beings making a slip or lapse are one of the most frequent forms of human error. This type of error is one accidentally committed by an experienced worker when undertaking standard tasks, for example, a worker may be fully aware of the importance of employing a safety harness, but one day, because distracted, they forget to put it on. Research has shown that for lapses to occur, the individual has to have their attention diverted elsewhere, either because they are preoccupied or distracted [62]. Human brains only have a limited capacity for paying attention [63]; if their attentional resources are diverted by a distraction, any other tasks undertaken simultaneously and demanding attention as well are adversely affected [64]. Thus, a distraction may undermine performance levels in the task at hand and result in adverse outcomes such as accident and/or injury [65, 66].

Because drones are still a new feature of the construction environment, they may be significantly distracting for workers on construction sites. Hearing or seeing a drone may distract a worker and interrupt their tasks. Because the interruption is unrelated to their task, it can provoke a cognitive shift and therefore disrupt the worker [67]. The worker may need as much as 25 minutes, having been distracted, before they are once again completely focused on their task [68]. In this period, workers may experience a drop in productivity and, more importantly, be more likely to make mistakes and endanger themselves and/or others [69]. reports that an interruption of just 2 seconds is sufficient to make a worker unfocused, and if a task is interrupted frequently, it results in workers being more stressed and frustrated [67]. A drone can be responsible for distracting workers visually, auditorily, and/or cognitively.

7.3.2.1 Visual Distractions

The model created by Cowen [70–72] proposes that a stimulus can cause distraction either because it is novel or because it is relevant. A novel though irrelevant stimulus can distract a worker's attention and impair their ability to process relevant data efficiently. On a construction site, seeing a drone represents a visual stimulus that is not constantly present resulting in worker focusing their attention toward drone and thus become distracted from their task. Because construction workers are frequently undertaking complex tasks, using dangerous equipment, in a hazardous and shifting work environment, they require the ability to be highly selective, and researchers have shown that the thoughts and actions directed toward a goal during these undertakings can experience significant disruption when distracted (a.k.a. goal neglect) [73, 74]. Because of this, a construction worker who is already experiencing exposure to a number of hazards has a greater likelihood of making mistakes or experiencing lapses as a result of visual distraction, and this has the potential to cause injury and/or fatality. Transportation researchers have already noted that small drones can be distracting to drivers in a way that could cause significant incidents [75, 76]. The same applies with construction, where those who work at height and who operate heavy machinery may experience distraction when drones fly above or nearby, resulting in the commission of unintentional hazardous actions or creating an unsafe environment that could cause injuries to themselves or others.

7.3.2.2 Auditory Distractions

With a visual stimulus, the worker has to make a choice to look at an object, but the sound comes from all directions, so that workers can hear a drone when in operation most of the time. When we hear additional sounds, they could be carrying important information for us, and so we by default devote mental capacity to processing the noise and evaluating whether it is relevant to us [77]. This is done by the brain without us devoting conscious thought to it or even before we realize the distraction is present, diverting a proportion of our attention away from the task we have in hand. Because of the way human hearing has evolved, we are highly skilled in reacting to any sound stimulus that could indicate approaching danger [78]. This means that we can diverse our attention subconsciously toward environmental noise even if we are trying to focus on another task. Although this can be favorable in some circumstances, it also means that we can become distracted by sounds that have no relevance to us. Those working in construction, in order to undertake their tasks safely, have to be selective and focused on the way they deploy their attention; irrelevant noise, e.g., that from a drone, can disrupt selective attention and distract from the task [78]. When we are distracted, cognitive performance is impaired and so lapses or errors may occur, resulting in an accident and/or injury [79]. Additionally, when we are distracted, we can feel emotionally disturbed/threatened [80], which can be a promoter of negative emotions.

7.3.2.3 Cognitive Distractions

Contrary to a popular belief, the human brain cannot multitask. We may feel that we are undertaking two tasks at the same time, but we are simply switching our attention back and forth between two or more tasks [81, 82]. If a construction worker is attempting to undertake a pair of tasks that both require a high level of recognition, e.g., undertake their construction tasks and at the same time think about the drone flying in their work environment or how it may be recording their activity and their brain that shifts attention between the two tasks. This could cause attention blindness [83], in the course of which a worker can be vulnerable to hazards they have not anticipated. Thus, even if the most obvious ways in which a drone can distract a worker (i.e., visually or auditorily) are not present, the simple knowledge that a drone can be somewhere on site may cognitively distract a worker. This is particularly the case if drones (e.g., safety drones or inspector drones) are employed for capturing and recording work quality for different work activities. Generally, workers can easily identify visual or auditory distractions and consciously decide to ignore them [84]. This form of cognitive distraction, however, is difficult to identify, and so the risks it creates could be longer-lasting.

There has been much discussion of the negative influence of distraction, especially in terms of safety. In the construction industry, Namian et al. [85] revealed that distraction in the workplace has an adverse impact on safety performance, perceptions of safety risks, and recognizing hazards. Researchers also demonstrated that drivers who become distracted have a greater likelihood of losing control of their vehicle, ignoring safety signs, and ultimately becoming involved in serious incidents [86]. In a similar way, medical practitioners commit serious diagnostic or treatment errors [87]. The manufacturing, aviation, and mining industries have also noted that distraction can have similar consequences [88]. For construction workers, the distraction caused by a drone may disturb the sensorimotor control mechanism and lead to a loss of balance, possibly increasing sway area or support base [89], leading to accidents caused by trips or falls. In addition, when several stimuli are present at the same time (e.g., drone stimulus and hazard stimulus), the worker can become cognitively overloaded [90]. Because of this, workers may find it difficult to make a rational quantification of risks as represented by other less obvious hazards [85]. Distraction can also impair a worker's situational awareness [91], which limits their capacity for proper assessment of the level of risk present, which can lead to unsafe behaviors and potential accidents.

7.3.3 Psychological Impact

Most drones carry visual sensors that are critical for its navigation and also take images or videos that can be employed in a number of ways, e.g., maintenance inspections, monitoring safety, tracking equipment and materials, or monitoring progress. Although the main intention of such drones is not to undertake worker monitoring, in the course of their primary activity, they inevitably record worker activity. Because of this, workers may start to feel that they are under constant observation, causing levels of anxiety and stress that could have an adverse impact on their mental health. Research by the National Institute of Occupational Safety and Health found that, in comparison with control groups, workers subjected to electronic monitoring had notably higher stress levels [92]. With drones, specific aspects of work can be monitors in real time, and non-task-related activity is also recorded. When workers know they are in such observation, their health may be adversely affected due to stress responses. Stress is not solely provoked by drones; rather it is caused by numerous physical and psychological factors existing in the environment [93]. Being persistently exposed to these environmental stresses over time can result in chronic biological/emotional arousal, which may lead to mental or psychophysiological illness [100]. This form of exposure can either engender pathological change or worsen an existing biological condition and ultimately result in workers developing chronic disease [94].

Additionally, if workers perceive that they are being watched, they may work faster and feel more pressured by time [95] as a result of the Hawthorne effect [96], which leads to a task's mental workload being increased. A number of researchers have demonstrated that there is a direct and consistent correlation between a worker's health and their mental workload [97, 98]. Monitoring (or believing they are experiencing monitoring) can engender stress as workers feel that the scenario (being monitored by a drone) is demanding something of them (not committing any errors) that may be beyond their capacity [99]. Concerns regarding making errors while under drone monitoring adds an extra cognitive demands to a task [100], resulting in side effects such as fatigue that makes accidents more likely [33, 101–104].

Lastly, being constantly distracted both visually and auditorily by a drone may result in sensory saturation/sensory overload [105], causing workers to experience negative emotions [106, 107]. Such emotions may include fatigue, problems concentrating, hyperhidrosis, and irritability [106]. Negative emotions caused by the presence of a drone on-site, or caused by distractions resulting from its presence, may provoke a number of physiological reactions. Research has demonstrated that the immune system can be altered by the emotional state of the individual [108]. Negative emotions set off a biological reaction within the hypothalamus-pituitary-adrenal (HPA) system [109] that impacts the immune system. This can then increase inflammatory responses, leading to disturbances in physiological equilibrium [110]. These responses are usually regarded as useful in coping with imminent dangers, but if they are frequently activated, these can be detrimental to worker health.

Individuals who are frequently exposed to such responses are more likely to suffer from depression and anxiety [111] which, in turn, means they are more likely to develop hypertension, type II diabetes, cardiovascular disease, or other health problems [112, 113]. The negative influence of these conditions can have a huge impact on a worker's personal and social lives, finances, and family life.

7.4 Future Research

Although this research has recognized and described a number of health and safety problems that may arise with the deployment of drones in the construction industry, all the described risk categories should be thoroughly investigated for the full understanding and evaluation of the problems that integrating drones could cause. Having investigated, new technologies and regulations should be created to make sure that construction workers are not negatively impacted in health and safety terms by the ever-greater presence of drones in construction. We propose a roadmap consisting of two branches pursued concurrently (Fig. 7.3). Firstly, all of the risk categories should be investigated and empirical studies undertaken to create an understanding and evaluation of the various ways drones can have an adverse influence on worker health and safety; secondly, the findings of these investigations should be used to create new regulations, protocols, and technology that allows drones to operate safely in the construction sector. The second branch can be pursued simultaneously with the first, with new information and discoveries being continuously exchanged. In addition, other important issues of human drone interaction such as trust, communication modalities, comfortability level, etc. should also be thoroughly evaluated.

7.5 Conclusion

The deployment of drones in construction is seeing a rapid increase. Although there has been discussion regarding the beneficial aspects of drones, there has been little research into the health and safety problems that may be caused by drones operating within construction sites. Drones can negatively influence the health and safety of workers through the introduction of unique risks of three types:

1. Physical risk, i.e., the likelihood of a drone or a part thereof coming into physical condition with a worker(s) or other parts of the construction site, e.g., equipment, structures. Such physical risks may be caused due to environmental factors or faulty software, sensors, or operational error.
2. Attentional cost, i.e., the risks that workers may become distracted due to seeing (visual distraction), hearing (auditory distraction), or simply knowing that they are in the presence of (cognitive distraction) drones.

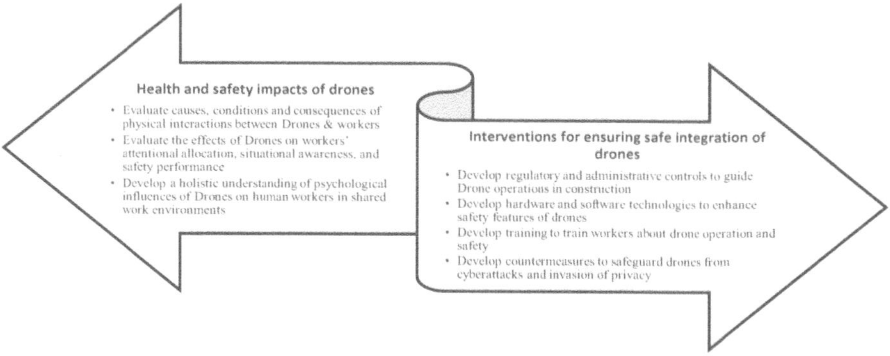

Fig. 7.3 Future research roadmap

3. Psychological impact, i.e., drones may negatively influence workers both physiologically and psychologically by causing sensory overload, cognitive overload, and high-stress levels.

Future research should explore different risk categories to improve the current state of knowledge regarding the ways in which humans and drones interact and to create measures that allow drones to be safely integrated into the construction industry. There is a need to undertake empirical evaluation of various elements of psychological impact, attentional costs, and physical risk, using experimentation for understanding and evaluation of the various means whereby worker health and safety may be adversely impacted by drones. We also need to use these findings to develop and evaluate new software and hardware for drones to promote safe operation of drones in the presence of human workers and to develop and evaluate regulations and protocols to ensure that drones can safely operate within the construction industry.

References

1. Grand View Research, Commercial Drone Market Worth $129.2 Billion By 2025 | CAGR: 56.5%. 2019. [Online]. Available: https://www.grandviewresearch.com/industry-analysis/global-commercial-drones-market
2. Drone Deploy, "The Rise of Drones in Construction," 2018
3. G. Albeaino, M. Gheisari, B.W. Franz, A systematic review of unmanned aerial vehicle application areas and technologies in the AEC domain. J. Inf. Technol. Constr. (2019). https://doi.org/10.36680/j.itcon.2019.020
4. S. Zhou, M. Gheisari, Unmanned aerial system applications in construction: A systematic review. Constr. Innov. (2018). https://doi.org/10.1108/CI-02-2018-0010

5. P. Liu et al., A review of rotorcraft unmanned aerial vehicle (drone) developments and applications in civil engineering. Smart Structures and Systems (2014). https://doi.org/10.12989/sss.2014.13.6.1065
6. M. Zucchii, "Drones: A Gateway Technology to Full Site Automation," 2015. Accessed: Jul. 06, 2020. [Online]. Available: https://www.enr.com/articles/9040-drones-a-gateway-technology-to-full-site-automation
7. Y. Ham, K.K. Han, J.J. Lin, M. Golparvar-Fard, Visual monitoring of civil infrastructure systems via camera-equipped unmanned aerial vehicles (drones): A review of related works. Visualization in Engineering (2016). https://doi.org/10.1186/s40327-015-0029-z
8. W.S. Hart, N.G. Gharaibeh, *Use of Micro Unmanned Aerial Vehicles in Roadside Condition Surveys* (2011). https://doi.org/10.1061/41167(398)9
9. U. Niethammer, M.R. James, S. Rothmund, J. Travelletti, M. Joswig, Drone-based remote sensing of the super-Sauze landslide: Evaluation and results. Eng. Geol. (2012). https://doi.org/10.1016/j.enggeo.2011.03.012
10. A. Koutsoudis, B. Vidmar, G. Ioannakis, F. Arnaoutoglou, G. Pavlidis, C. Chamzas, Multi-image 3D reconstruction data evaluation. J. Cult. Herit. (2014). https://doi.org/10.1016/j.culher.2012.12.003
11. M. Uysal, A.S. Toprak, N. Polat, Photo realistic 3d modeling with drone: Gedik Ahmet pasha mosque in Afyonkarahisar. ISPRS - Int. Arch. Photogramm. Remote Sens. Spat. Inf. Sci (2013). https://doi.org/10.5194/isprsarchives-xl-5-w2-659-2013
12. A. Banaszek, A. Zarnowski, A. Cellmer, S. Banaszek, Application of new technology data acquisition using aerial (drone) digital images for the needs of urban revitalization. Environ. Eng. (2017)
13. D. Bulatov, P. Solbrig, H. Gross, P. Wernerus, E. Repasi, C. Heipke, Context-based urban terrain reconstruction from drone-videos for geoinformation applications. ISPRS - Int. Arch. Photogramm. Remote Sens. Spat. Inf. Sci (2012). https://doi.org/10.5194/isprsarchives-xxxviii-1-c22-75-2011
14. G.M. Waehrer, X.S. Dong, T. Miller, E. Haile, Y. Men, Costs of occupational injuries in construction in the United States. Accid. Anal. Prev. **39**(6), 1258–1266 (2007). https://doi.org/10.1016/j.aap.2007.03.012
15. Bureau of Labor Statistics, "Census of Fatal Occupational Injuries (CFOI) - Current and Revised Data," 2019
16. Bureau of Labor Statistics, "Fatal Injury Report 2019," 2019. [Online]. Available: https://www.bls.gov/iif/oshwc/cfoi/cftb0324.htm
17. M. Gheisari, B. Esmaeili, Applications and requirements of unmanned aerial systems (UASs) for construction safety. Saf. Sci. **118**, 230–240 (2019)
18. T. Wang, "U.S. Construction Industry - Statistics & Facts," 2019. doi: https://www.statista.com/topics/974/construction/
19. T. Wang, "Value added by the construction industry as a share of gross domestic product in the U.S. from 2007 to 2019," 2020
20. Bureau of Labor Statistics, "Injury Report 2017," 2017. [Online]. Available: https://www.bls.gov/iif/oshcfoi1.htm
21. Bureau of Labor Statistics, "Occupational Employment Statistics," 2020. [Online]. Available: https://www.bls.gov/oes/current/oes_nat.htm#47-0000
22. P. X. W. Zou and R. Y. Sunindijo, Strategic Safety Management in Construction and Engineering. 2015
23. A. Albert, M.R. Hallowell, B.M. Kleiner, Enhancing construction Hazard recognition and communication with energy-based cognitive mnemonics and safety meeting maturity model: Multiple baseline study. J. Constr. Eng. Manag. **140**(2), 04013042 (2014). https://doi.org/10.1061/(ASCE)CO.1943-7862.0000790
24. J. W. Mroszczyk, "Improving Construction Safety: a Team Effort," Prof. Saf., pp. 55–68, 2015, [Online]. Available: https://aeasseincludes.assp.org/professionalsafety/pastissues/060/06/F2Mros_0615.pdf

25. S.D. Choi, L. Guo, J. Kim, S. Xiong, Comparison of fatal occupational injuries in construction industry in the United States, South Korea, and China. Int. J. Ind. Ergon. (2019). https://doi.org/10.1016/j.ergon.2019.02.011
26. CPWR–The Center for Construction, The construction chart book, 4th ed. Cpwr-The Center for Construction Research and Training, 2007
27. CPWR–The Center for Construction, The U.S. Construction Industry and Its Workers, Sixth. 2018
28. R.A. Haslam et al., Contributing factors in construction accidents. Appl. Ergon. **36**(4), 401–415 (2005). https://doi.org/10.1016/j.apergo.2004.12.002
29. S. Winge, E. Albrechtsen, B.A. Mostue, Causal factors and connections in construction accidents. Saf. Sci. (2019). https://doi.org/10.1016/j.ssci.2018.10.015
30. A. Suraji, A.R. Duff, S.J. Peckitt, Development of causal model of construction accident causation. J. Constr. Eng. Manag. (2001). https://doi.org/10.1061/(ASCE)0733-9364(2001)127:4(337)
31. O.S. Williams, R. Adul Hamid, M.S. Misnan, Accident causal factors on the building construction sites: A review. Int. J. Built Environ. Sustain. (2018). https://doi.org/10.11113/ijbes.v5.n1.248
32. C. Perrow, Normal Accidents : Living with High Risk Technologies Updated Edition. (1999)
33. L. Cavuoto and F. Megahed, "Understanding Fatigue and the Implications for Worker Safety," ASSE Professional Development Conference and Exposition. American Society of Safety Engineers, Atlanta, Georgia, USA, p. 10, 2016, [Online]. Available: https://doi.org/
34. M. Hassanalian, A. Abdelkefi, Classifications, applications, and design challenges of drones: A review. Prog. Aerosp. Sci. (2017). https://doi.org/10.1016/j.paerosci.2017.04.003
35. Drone Deploy, "2018 Commercial Drone Industry Trends," 2018. [Online]. Available: https://www.dronedeploy.com/resources/ebooks/2018-commercial-drone-industry-trends-report/#form
36. J.S. Park, P. Kim, Y.K. Cho, Y. Fang, Automated collaboration framework of drone and UGV for 3D visualization of construction sites. 18th Int. Conf. Constr. Appl. Virtual Real. (2018)
37. S. Siebert, J. Teizer, Mobile 3D mapping for surveying earthwork projects using an unmanned aerial vehicle (drone) system. Autom. Constr. **41**, 1–14 (2014). https://doi.org/10.1016/j.autcon.2014.01.004
38. D. Kang, Y.J. Cha, Autonomous drones for structural health monitoring using deep learning and an ultrasonic Beacon system with geo-tagging. Comput. Civ. Infrastruct. Eng (2018). https://doi.org/10.1111/mice.12375
39. J.G. Martinez, M. Gheisari, L.F. Alarcón, Drone integration in current construction safety planning and monitoring processes: Case study of a high-rise building construction project in Chile. J. Manag. Eng. (2020). https://doi.org/10.1061/(ASCE)ME.1943-5479.0000761
40. Q. Lindsey, D. Mellinger, V. Kumar, *Construction of Cubic Structures with Quadrotor Teams* (2012). https://doi.org/10.15607/rss.2011.vii.025
41. S. Schneegass, F. Alt, J. Scheible, A. Schmidt, *Midair Displays: Concept and First Experiences with Free-Floating Pervasive Displays* (2014). https://doi.org/10.1145/2611009.2611013
42. S. Goessens, C. Mueller, P. Latteur, Feasibility study for drone-based masonry construction of real-scale structures, in *Automation in Construction*, (2018). https://doi.org/10.1016/j.autcon.2018.06.015
43. A. Bürkle, F. Segor, M. Kollmann, *Towards Autonomous Micro Drone Swarms* (2011). https://doi.org/10.1007/s10846-010-9492-x
44. R. Melo, D. Costa, *Reducing the Gap between Work as Done and Work as Imagined on Construction Safety Supported by UAS* (2020). https://doi.org/10.15626/rea8.02
45. M. Gheisari, A. Rashidi, B. Esmaeili, Using unmanned aerial systems for automated fall hazard monitoring, in *Proceedigs of ASCE Construction Research Congress*, (2018), pp. 62–72
46. R.R.S. de Melo, D.B. Costa, J.S. Álvares, J. Irizarry, Applicability of unmanned aerial system (UAS) for safety inspection on construction sites. Saf. Sci. (2017). https://doi.org/10.1016/j.ssci.2017.06.008

47. S. Kim, J. Irizarry, D.B. Costa, *Potential Factors Influencing the Performance of Unmanned Aerial System (UAS) Integrated Safety Control for Construction Worksites* (2016). https://doi.org/10.1061/9780784479827.260
48. J. Irizarry, D.B. Costa, Exploratory study of potential applications of unmanned aerial Systems for Construction Management Tasks. J. Manag. Eng. (2016). https://doi.org/10.1061/(ASCE)ME.1943-5479.0000422
49. R.R.S. de Melo, D.B. Costa, Integrating resilience engineering and UAS technology into construction safety planning and control. Eng. Constr. Archit. Manag. (2019). https://doi.org/10.1108/ECAM-12-2018-0541
50. S. Alizadehsalehi, I. Yitmen, T. Celik, D. Arditi, The effectiveness of an integrated BIM/drone model in managing safety on construction sites. Int. J. Occup. Saf. Ergon., 1–16 (2018)
51. S. Kim, J. Irizarry, D.B. Costa, Field test-based UAS operational procedures and considerations for construction safety management: A qualitative exploratory study. Int. J. Civ. Eng (2020). https://doi.org/10.1007/s40999-020-00512-9
52. C.M. Mendes, B.F. Silveira, D.B. Costa, R.S. Melo, Evaluating USA-image pattern recognition system application for safety guardrails inspection, in *Proceedings of the Joint CIB W099 and TG59 Conference Coping with the Complexity of Safety, Health, and Wellbeing in Construction. Salvador, Brazil*, (2018), pp. 23–32
53. D. Roberts, T. Bretl, M. Golparvar-Fard, *Detecting and Classifying Cranes Using Camera-Equipped Drones for Monitoring Crane-Related Safety Hazards* (2017). https://doi.org/10.1061/9780784480847.055
54. M. Gheisari, J. Irizarry, B.N. Walker, *UAS4SAFETY: The potential of Unmanned Aerial Systems for Construction SAFETY Applications* (2014). https://doi.org/10.1061/9780784413517.0184
55. J. Irizarry, M. Gheisari, B.N. Walker, Usability assessment of drone technology as safety inspection tools. Electron. J. Inf. Technol. Constr. (2012)
56. M. Vasic, A. Billard, Safety issues in human-robot interactions, in *2013 Ieee International Conference on Robotics and Automation*, (2013), pp. 197–204
57. U.S. Occupational Safety and Health Administration (OSHA), Chapter 6 – Robotics in the workplace, in *Machine Safeguarding*, (2019)
58. A.J. Kerns, D.P. Shepard, J.A. Bhatti, T.E. Humphreys, Unmanned aircraft capture and control via GPS spoofing. J. F. Robot (2014). https://doi.org/10.1002/rob.21513
59. Mercury News, "Mountain View: Drone crash knocks out power to 1,600," Mercury News, Mountain View, Jun. 09, 2017
60. BBC News, *Drone Crash Causes Hollywood Electricity Blackout - BBC News* (BBC News, Hollywood, 2015)
61. D. Kasper, A. Fauci, S. Hauser, D. Longo, J. Jameson, and J. Loscalzo, Harrison's principles of internal medicine, 19e, vol. 1, no. 2. Mcgraw-hill, 2015
62. N. Meshkati, Safety and human factors considerations in control rooms of oil and gas pipeline systems: Conceptual issues and practical observations. Int. J. Occup. Saf. Ergon. (2006). https://doi.org/10.1080/10803548.2006.11076669
63. D. Kahneman, *Attention and Effort*, vol 1063 (Citeseer, 1973)
64. R.W. Weisberg, L.M. Reeves, *Cognition.: From Memory to Creativity* (Wiley, 2013)
65. S. Berti, E. Schröger, A comparison of auditory and visual distraction effects: Behavioral and event-related indices. Cogn. Brain Res. (2001). https://doi.org/10.1016/S0926-6410(00)00044-6
66. S. Berti, E. Schröger, Distraction effects in vision: Behavioral and event-related potential indices. Neuroreport **15**(4), 665–669 (2004)
67. G. Mark, D. Gudith, U. Klocke, *The Cost of Interrupted Work: More Speed and Stress* (2008). https://doi.org/10.1145/1357054.1357072
68. G. Mark, V.M. Gonzalez, J. Harris, No task left behind? Examining the nature of fragmented work, in *Proceedings of the SIGCHI Conference on Human Factors in Computing Systems*, (2005), pp. 321–330

69. E.M. Altmann, J.G. Trafton, D.Z. Hambrick, *Momentary Interruptions Can Derail the Train of Thought* (J. Exp. Psychol. Gen, 2014). https://doi.org/10.1037/a0030986
70. N. Cowan, An embedded-processes model of working memory, in *Models of Working Memory*, (2012)
71. N. Cowan, Evolving conceptions of memory storage, selective attention, and their mutual constraints within the human information-processing system. Psychol. Bull. (1988). https://doi.org/10.1037/0033-2909.104.2.163
72. N. Cowan, *Attention and Memory: An Integrated Framework* (Oxford University Press, 1999)
73. J. Duncan, H. Emslie, P. Williams, R. Johnson, C. Freer, Intelligence and the frontal lobe: The organization of goal-directed behavior. Cogn. Psychol. (1996). https://doi.org/10.1006/cogp.1996.0008
74. R. West, Visual distraction, working memory, and aging. Mem. Cogn. (1999). https://doi.org/10.3758/BF03201235
75. A. M. Ryan, "Driver Performance Due to Unmanned Aerial System Applications in the Vicinity of Surface Transportation," 2019
76. D. S. Hurwitz, M. J. Olsen, and Z. Barlow, "Driving Distraction Due to Drones," Oregon. Dept. of Transportation. Research Section, 2018
77. E. Schröger, M.H. Giard, C. Wolff, Auditory distraction: Event-related potential and behavioral indices. Clin. Neurophysiol. (2000). https://doi.org/10.1016/S1388-2457(00)00337-0
78. S.P. Banbury, W.J. Macken, S. Tremblay, D.M. Jones, Auditory distraction and short-term memory: Phenomena and practical implications. Hum. Factors (2001). https://doi.org/10.1518/001872001775992462
79. D. E. Broadbent, Decision and Stress. 1971
80. J. William, *The Principles of Psychology*, vol 1 (1890)
81. J.M. McDowd, An overview of attention: Behavior and brain. J. Neurol. Phys. Ther. **31**(3), 98–103 (2007)
82. C.M. Dzubak, Multitasking: The good, the bad, and the unknown. J. Assoc. Tutoring Prof. **1**(2), 1–12 (2008)
83. A. Mack, I. Rock, *Inattentional blindness* (MIT press, 1998)
84. National Safety Council, Understanding the distracted brain: Why driving while using hands-free cell phones is risky behavior. 2012, [Online]. Available: https://www.nsc.org/Portals/0/Documents/DistractedDrivingDocuments/Cognitive-Distraction-White-Paper.pdf
85. N. Mostafa, A. Alex, F. Jing, Effect of distraction on Hazard recognition and safety risk perception. J. Constr. Eng. Manag. **144**(4), 4018008 (2018). https://doi.org/10.1061/(ASCE)CO.1943-7862.0001459
86. K.L. Young, P.M. Salmon, Examining the relationship between driver distraction and driving errors: A discussion of theory, studies and methods. Saf. Sci. **50**(2), 165–174 (2012)
87. A.J. Rivera-Rodriguez, B.-T. Karsh, Interruptions and distractions in healthcare: Review and reappraisal. BMJ Qual. Saf. **19**(4), 304–312 (2010)
88. Y.-H. Chang, H.-H. Yang, Y.-J. Hsiao, Human risk factors associated with pilots in runway excursions. Accid. Anal. Prev. **94**, 227–237 (2016)
89. B.G. Rasman, P.A. Forbes, R. Tisserand, J.S. Blouin, *Sensorimotor Manipulations of the Balance Control Loop-beyond Imposed External Perturbations* (Front. Neurol, 2018). https://doi.org/10.3389/fneur.2018.00899
90. D.E. Meyer, D.E. Kieras, *A Computational Theory of Executive Cognitive Processes and Multiple-Task Performance: Part 1. Basic Mechanisms* (Psychol. Rev, 1997). https://doi.org/10.1037/0033-295X.104.1.3
91. D.L. Strayer, D.L. Fisher, SPIDER: A framework for understanding driver distraction. Hum. Factors **58**(1), 5–12 (2016)
92. health and human services Department of Labor, Departments of Labor, Health and Human Services, Education, and Related Agencies Appropriations for 1993: Department of Health and Human Services, January 1. U.S. Government Printing Office, 1992

93. B.C. Amick III, M.J. Smith, Stress, computer-based work monitoring and measurement systems: A conceptual overview. Appl. Ergon. **23**(1), 6–16 (1992)
94. P.L. Schnall, J.E. Schwartz, P.A. Landsbergis, K. Warren, T.G. Pickering, *Relation between Job Strain, Alcohol, and Ambulatory Blood Pressure* (Hypertension, 1992). https://doi.org/10.1161/01.HYP.19.5.488
95. S. Sarpong, D. Rees, *Assessing the Effects of 'Big Brother' in a Workplace: The Case of WAST* (Eur. Manag. J, 2014). https://doi.org/10.1016/j.emj.2013.06.008
96. R. McCarney, J. Warner, S. Iliffe, R. Van Haselen, M. Griffin, P. Fisher, The Hawthorne effect: A randomised, controlled trial. BMC Med. Res. Methodol. **7**(1), 30 (2007)
97. T. Theorell, B. Floderus-Myrhed, 'Workload'and risk of myocardial infarction—A prospective psychosocial analysis. Int. J. Epidemiol. **6**(1), 17–21 (1977)
98. J.S. House, *Occupational Stress and the Mental and Physical Health of Factory Workers*, Research Report Series (ERIC, 1980)
99. J.E. McGrath, Stress and behavior in organizations: A handbook of industrial and organizational psychology. Chicago Rand McNally. Menag. E.(1982). Meas. coping Eff. A panel Anal. Marital Probl. coping efforts. J. Heal. Soc. Behav. **23**, 220–234 (1976)
100. T. Van Gog, L. Kester, F. Paas, Effects of concurrent monitoring on cognitive load and performance as a function of task complexity. Appl. Cogn. Psychol. (2011). https://doi.org/10.1002/acp.1726
101. A. Williamson, D.A. Lombardi, S. Folkard, J. Stutts, T.K. Courtney, J.L. Connor, The link between fatigue and safety. Accid. Anal. Prev. (2011). https://doi.org/10.1016/j.aap.2009.11.011
102. D. Dawson, A. Fletcher, *A Quantitative Model of Work-Related Fatigue: Background and Definition* (Ergonomics, 2001). https://doi.org/10.1080/00140130119399
103. H. Hsiao, P. Simeonov, *Preventing Falls from Roofs: A Critical Review* (Ergonomics, 2001). https://doi.org/10.1080/00140130110034480
104. M. Chan, Fatigue: The most critical accident risk in oil and gas construction. Constr. Manag. Econ. (2011). https://doi.org/10.1080/01446193.2010.545993
105. G. Simmel, The metropolis and mental life, in *The Urban Sociology Reader*, (Routledge, 2012), pp. 37–45
106. Z.J. Lipowski, Sensory and information inputs overload: Behavioral effects. Compr. Psychiatry (1975)
107. L.A. Gottschalk, J.L. Haer, D.E. Bates, Effect of sensory overload on psychological state: Changes in social alienation-personal disorganization and cognitive-intellectual impairment. Arch. Gen. Psychiatry (1972). https://doi.org/10.1001/archpsyc.1972.01750280019004
108. J.W. Kalat, *Biological Psychology* (Nelson Education, 2015)
109. N. Michels et al., Negative life events, emotions and psychological difficulties as determinants of salivary cortisol in Belgian primary school children. Psychoneuroendocrinology **37**(9), 1506–1515 (2012)
110. F. D'Acquisto, L. Rattazzi, G. Piras, Smile—It's in your blood! Biochem. Pharmacol. **91**(3), 287–292 (2014)
111. C. Kubo, Stress and depression. Nippon rinsho. Japanese journal of clinical medicine (2007). https://doi.org/10.1146/annurev.clinpsy.1.102803.143938
112. B.E. Cohen, D. Edmondson, I.M. Kronish, State of the art review: Depression, stress, anxiety, and cardiovascular disease. Am. J. Hypertens. **28**(11), 1295–1302 (2015)
113. T.G. Pickering, Mental stress as a causal factor in the development of hypertension and cardiovascular disease. Curr. Hypertens. Rep. **3**(3), 249–254 (2001)

Chapter 8
Assessing the Role of Wearable Robotics in the Construction Industry: Potential Safety Benefits, Opportunities, and Implementation Barriers

Ifeanyi Okpala, Chukwuma Nnaji, Omobolanle Ogunseiju, and Abiola Akanmu

Abstract High rate of work-related musculoskeletal disorders (WMSDs) remains a major concern of practitioners and researchers in the construction industry. Among current developments in automation and robotics, wearable robots (exoskeletons) have gained some traction as a viable tool that could improve construction workers' safety, health, and productivity at work. However, limited studies have explored the efficacy and effectiveness of existing exoskeletons in reducing risks of WMSD in the construction domain. Also, information on factors that prevent the successful implementation of wearable robots on construction projects is scarce. To bridge this gap, the present study presents the result of a multiphase process aimed at providing an in-depth perception-based evaluation of the capability of commercially available wearable robots. This process comprises of a detailed literature review, insights from an online survey, and usability studies from a lab-simulated construction task. First, 11 wearable robots with potential utility in construction operations were identified followed by a detailed assessment of the potential role these technologies could play in reducing well-established risk factors of work-related musculoskeletal injuries. Also, the authors evaluated the suitability of these robots for construction work involving heavy or repetitive lifting, and prolonged stooped postures, and synthesized the barriers to the acceptance of these robots. Results from the present study indicate that wearable robots could prevent about 60% of construction-related WMSDs impacting the different body regions (lower and upper back, feet and

I. Okpala (✉) · C. Nnaji
Department of Civil, Construction and Environmental Engineering, The University of Alabama, Tuscaloosa, AL, USA
e-mail: iuokpala@crimson.ua.edu; cnnaji@eng.ua.edu

O. Ogunseiju · A. Akanmu
Myers-Lawson School of Construction, Virginia Tech, Blacksburg, VA, USA
e-mail: omobolanle@vt.edu; abiola@vt.edu

H. Jebelli et al. (eds.), *Automation and Robotics in the Architecture, Engineering, and Construction Industry*, https://doi.org/10.1007/978-3-030-77163-8_8

ankle, wrist and hand, and shoulder) and 30–40% of accidents associated with critical WMSD risk factors.

Keywords Wearable robotics · Construction technology · Safety performance · Ergonomic exposures · Work-related musculoskeletal disorders · Automation

8.1 Introduction

The highly hazardous and physically demanding nature of construction activities contributes significantly to the disproportionate high rate of work-related musculoskeletal disorders (WMSDs) among construction workers [1, 2]. According to the Bureau of Labor Statistics [3], construction workers in the United States have an 18% higher rate of WMSDs when compared to all private industries when combined. From this BLS data, the body parts most affected are the shoulder and back, thus accounting for 14.8% and 36.9% of the total number of industry cases, respectively [3]. The foregoing is caused by the highly repetitive and complex nature of construction work which typically involves non-optimal postures, inherently irregular work times, variable work activities, and manual material or task handling [1, 4]. Despite existing and emerging injury prevention strategies (including job hazard analysis, incentives, employee training, record keeping, accident analysis, and frequent worksite inspections) being implemented on sites, the current occupational injury and illness situation subsists [5]. This has prompted researchers and practitioners to explore existing and novel technological innovations to improve workers' safety performance and productivity with perceived accompanied project cost and schedule gains [2].

It is well known that technology makes work easier. Several construction technologies are currently being investigated in a bid to enhance worker safety during construction work. As highlighted by [6], these technologies include artificial intelligence, building information modeling, immersive technologies, Internet of things and sensing technologies, smart camera systems, robotics and automation, photogrammetry and LiDAR, and unmanned aerial vehicles. Among these tools, [7] showcased that robotics and automation has emerged as a very promising frontier due to current advances in devices' development and promising personalized workers' use as opposed to traditional construction use cases (processing raw materials and building parts' production). Robotics and automation have been successfully employed in the prevention of occupational injuries and improvement of workers' productivity in health and social care [8], manufacturing [9], and automotive [10] industries. Among current developments in automation and robotics, wearable robotics (exoskeletons) have gained some traction as a viable tool that can improve construction worker safety, health, and productivity at work [1, 4, 7].

Wearable robots are highly reputed as an emerging and innovative intervention opportunity given its perceived application benefits such as augmentation, assistance, or/and enhancement of workers' physical activity by reducing physical demands and fatigue [1]. [8] opines that it is relatively easy to apply exoskeletons to construction work because of movement support (bending down or lifting heavy load) and prevention of repetitive injuries. However, research suggests that the adoption and use of wearable technology within construction are not widespread [11]. The authors have also determined that a limited number of studies have explored the potential efficacy and effectiveness of existing exoskeletons to determine applicability within the construction domain and potential to reduce risks of WMSD in the construction industry. To provide this, this chapter provides (1) an in-depth literature review on the use of exoskeleton in construction, (2) an assessment of the ability of wearable robots to prevent WMSDs impacting body parts and WMSDs' risk factors via an online survey of construction practitioners, and (3) an assessment of perceived exoskeleton usability during designated construction tasks. Insights summarized in this book chapter could assist in solving wearable robot integration problems in practice and inform future research directions on development and application of wearable robots in construction.

8.2 Background

8.2.1 Work-Related Musculoskeletal Disorders (WMSDs) in Construction

According to the US Bureau of Labor Statistics (BLS) [12], work-related musculoskeletal disorders (WMSDs) include:

> "an injury or illness from pinched nerve; herniated disc; meniscus tear; sprains, strains, and tears; hernia (traumatic and non-traumatic); pain, swelling, and numbness; carpal or tarsal tunnel syndrome; Raynaud's syndrome or phenomenon; and musculoskeletal system and connective tissue diseases and disorders..."

According to *The Construction Chart Book* released by the Center for Construction Research and Training (CPWR) in [13], the number of cases of WMSDs in construction (20,510) has dropped by 40% from 1992 up until the end of year 2015. The May 2020 BLS factsheet shows that during the year 2018, the construction industry recorded 19,380 [14]. Although there has been a significant reduction of WMSD case numbers and rates, the current situation remains a cause of concern. For instance, WMSDs cost companies over $3 billion annually [15]. Therefore, researchers have closely examined the risk factors associated with WMSD occurrence and the body regions of occurrence. As reported by [16, 17], the distinct human body regions are (1) shoulder and upper arm, (2) elbow, (3) lower arm and wrist, (4) hand, (5) hip, (6) upper leg, (7) knee, (8) lower leg and ankle, (9) foot, (10) neck, and (11) back. These regions can be considered singularly or in

Table 8.1 Risk factors for musculoskeletal disorders [9]

S./ no.	Risk factors for hands, wrist, neck, and shoulders	Risk factors for back injuries and pain	Other risk factors
1	Forcefulness or muscle effort	Lifting	Height of the work
2	Awkward body posture	Pushing, pulling, tugging	Working beyond your capacity
3	Repetitive work	Twisting, reaching, sideways bending, unequal lifting	Floor surfaces and obstacles in the work area and housekeeping practices
4	Vibration from hand tools	Working in a single position	Lifting techniques
5	External contact stress from tools and sharp objects	Whole-body vibration	Tool belts
6	-	-	Temperature

groups when attributing or examining risk factors which inherently lead to WMSDs. Table 8.1 presents a comprehensive list of risk factors for WMSDs resulting in days away from work, as reported by [18] in the electronic library of construction occupational safety and health. When examining injuries attributed to WMSDs in year 2015, sprains, strains, and tears (SST) leads the chart (65%) followed by soreness and pain (18%) and others (17%). Factors within the "risk factors for hands, wrist, neck, and shoulders" in Table 8.1 (first column) are factors of greatest concern because they are more likely to cause SSTs.

Among these risk factors, [13] showcased that overexertion (from lifting and lowering), overexertion (except lifting, pushing and pulling), and exertion (from bending to crawling) are top factors for WSMDs leading to lost workdays with 29.9 %, 16.4 %, and 11.7 % of the total 2015 cases, respectively. In 2015, the top 5 high-risk construction trades (based on exertion injuries per 10,000 full-time employees) are finish carpentry (56.1), tile and terrazzo (56.1), plumbing and heat A/C (46.5), drywall and insulation (45.1), and roofing (39.2). Implementing solutions, such as emerging technologies, that could address these unfavorable statistics is paramount.

8.2.2 *Wearable Robotics (Exoskeletons) in Construction*

Wearable robotics or exoskeletons are mechanical devices physically attached or worn by the construction worker to enhance human performance as opposed to a robot autonomously performing tasks instead of the worker [1, 11]. Commercially available exoskeletons exist largely in two broad forms: devices that inform wearers of the risk of overexertion or unsafe postures and others that reduce the impact of physical tasks (lifting heavy loads and using applicable tools) on construction workers. As further explained by [19], there are two broad categories of exoskeletons, namely, lower-limb and upper-limb exoskeletons. This implies that one category of

Fig. 8.1a A FLx ErgoSkeleton: back and side views

exoskeletons assists upper-limb movements (reaching, grasping, or lifting), while the lower-limb category assists functions such as sitting down, standing up, walking, balancing, and squatting. A wearable robotic system which combines the functions of both is designated as a full-body exoskeleton. Figure 8.1a is an example of the FLx ErgoSkeleton, an upper-body (back support) exoskeleton that focuses exclusively on keeping the position of the workers' body within a safe body posture while lifting or moving heavy objects. Figure 8.1b depicts a full-body exoskeleton (Flex Lift) that supports the knees and back and consequently encourages safe lifting, standing, and squatting postures. Aside from the potential utility, the choice of the FLx ErgoSkeleton device was made due to availability and willingness of StrongArm Technologies to facilitate procurement for testing.

Also, wearable robots can either be active (powered) or passive. The active exoskeletons have actuators and a power source that greatly augments workers' strength and endurance [20], while passive exoskeletons rely solely on springs and dampers to help reduce workers' injuries and to maintain healthier workforce systems [21]. According to [11], it is very important to note that exoskeletons could be a viable solution to the challenges the construction industry battles by having an aging workforce. This device will enable older construction workers to cope better with physically demanding tasks, leading to better worker retention.

When considering the use of exoskeletons in construction practice, some implementation barriers as reported by [1, 11] exist. As highlighted in Table 8.2, these barriers include potential introduction of safety and health risks, lack of technology integration, usability concerns, and exoskeleton technology adoption anxieties.

Fig. 8.1b Kinetic Edge Flex Lift: front and back views

Table 8.2 Implementation barriers to exoskeleton use

S./ no.	Barriers	Segments
1	Safety and health concerns	Catch and snag risks Falls False sense of safety Allergic reaction and hygiene issues
2	Usability concerns	Durability and ruggedness Versatility Effectiveness Peer acceptance and opinion Comfort
3	Lack of integration	Compatibility with other personal protective equipment (PPE)
4	Financial gain	Cost-benefit Return on investment Training cost

In response to these perceived barriers, several research studies funded by renowned bodies such as the Center for Construction Research and Training (CPWR), Construction Industry Institute (CII), National Science Foundation (NSF), and National Institute for Occupational Safety and Health (NIOSH) are currently ongoing. To assist these and future research efforts, it is necessary to identify commercially available exoskeletons that are applicable to the construction domain and also gauge the opinions of construction practitioners on the current use and perceived effectiveness of these exoskeletons.

8.3 Research Methodology

To achieve the research goal, the authors implemented a three-step research process (Fig. 8.2). First, a literature review was conducted to determine the categories of wearable robots and implementation barriers and identify risk factors and body regions associated with WMSDs in construction. Also, wearable robots with potential utility in construction operations were determined. Second, an online survey was created and distributed to the construction practitioners purposively sampled from the US construction industry. The survey consisted of two sections. The first section asked about background information (such as personal attributes such as role, job title, type of company/organization, and years of experience) of respondents. The second section requested the respondents to (1) indicate their familiarity with the categories of exoskeletons in the construction industry using a 5-point Likert rating scale (i.e., 1, Not at all familiar; 2, Slightly familiar; 3, Somewhat familiar; 4, Moderately familiar; 5, Extremely familiar); (2) gauge their perception regarding the potential of wearable robots to prevent WMSDs associated with established body regions (1 = No, 2 = Yes); and (3) gauge their views on the ability of exoskeletons to prevent accidents (0 = will not prevent, 100 = will prevent most accidents) associated with highlighted WMSDs' risk factors. Finally, ten participants were recruited to assess the performance of the Flex Lift exoskeleton when used to perform laboratory-simulated flooring task. After performing the required task, participants completed a questionnaire on the usability of the exoskeleton. The questionnaire for the assessment of exoskeleton usability was designed on a 5-point Likert scale which varied from 1= Strongly Disagree to 5= Strongly Agree. The usability questions were based on three criteria: ease of use, ease of learning, and comfort.

Fig. 8.2 Research methodology

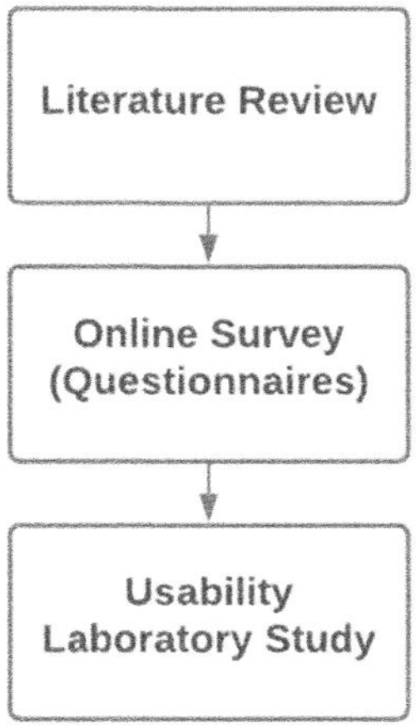

8.4 Results and Discussions

8.4.1 Existing Exoskeleton Applications in Construction

To identify existing exoskeletons applicable in construction, the authors studied industrial reports and research papers covering research, development, and the applicability of industrial exoskeletons in both construction and other industries including automotive and healthcare. These studies [19, 21, 22] were complemented by insights from the educational catalog on exoskeletonreport.com to provide a holistic overview of the exoskeleton industry. This process also provided the authors with a method to sort exoskeletons by type, purpose, and promise [24]. Table 8.3 shows 11 wearable robots with potential utility in construction operations, associated tasks, characteristics, and categories.

From the exoskeletons identified, majority are passive, thus indicating that manufacturers believe that devices considered heavy or requiring periodic electrical charging may not find widespread acceptance in the construction space. Also, when looking at the characteristics of the wearable robots, the list is dominated by upper-body robots, thus showing that the industry is lacking portable and comfortable lower-body devices. One potential reason for the skewed preference for upper-body devices is the fact that WMSDs in the construction industry occur mainly in the shoulder and back regions [3]. Few technologies, such as FORTIS, develop products that take into consideration the need for construction workers to efficiently move from one place to another. In most cases, manufacturers of the listed exoskeletons have experimentally adjudged them to be capable of leading to improvements

Table 8.3 Identified exoskeletons applicable in construction

S./no.	Wearable robot	Tasks	Characteristic	Category
1	FLx ErgoSkeleton	Pick and carry	Upper body	Passive
2	Hilti EXO-01	Elevated arms and repetitive arm motions	Upper body	Passive
3	SuitX MAX	Bending, lifting, squatting, elevated arms, and prolonged standing	Full body	Passive
4	Flex Lift	Bending, lifting, and standing	Full body	Passive
5	EksoVest	Elevated arms, static arms, and repeated arm motions	Upper body	Passive
6	Iron Hand	Grasping	Upper body (Wrist)	Active
7	FORTIS	Heavy tool and static arms	Semi-full body	Passive
8	ATOUN Model Y	Bending and lifting	Upper body	Active
9	Levitate Airframe	Elevated arms, static arms, and repeated arm motions	Upper body	Passive
10	Laevo	Bending, lifting, and reaching	Upper body	Passive
11	backX	Bending, lifting, reaching, and stooping	Upper body	Passive

in task productivity and workers' safety. However, there is a need to assess these technologies in the construction industry and recommend quantifiable criteria for developing and configuring exoskeleton technology which is amenable to different construction trades. Providing additional information specific to the construction industry will clearly identify limitations and solve performance-related issues [22].

8.4.2 Online Survey Respondent Characteristics

Fifty-one construction practitioners, aware of the specific safety management practices and technologies implemented by their organization, were sampled. Of this number, 84.31 % of these professionals have at least a bachelor's degree with at least 62% indicating that they are very highly proficient in project management and construction safety. Approximately 90% of respondents are in upper management roles, which indicate that the data is reliable since the participants are very knowledgeable. Table 8.4 presents a summary of the profile of the respondents. Moreover, considering that the pool contains management-level staff with high construction experience (all participants have spent at least 5 years on the job), individual perception will reflect decision-making experience in addition to the knowledge acquired from planning, designing, and executing injury prevention strategies and other safety programs. The design of the usability questionnaire was guided by the system usability scale [23] and similar studies such as [22, 25]. The questions were designed to capture users' perceptions of the main usability characteristics proposed by [25], e.g., comfort, interference, simplicity, satisfaction, relevance, and acceptance of the exoskeleton.

Table 8.4 Profile of Respondents (Construction Practitioners)

Field	Count	Percentage
Source of the respondent's organization		
General contractor	49	96.08%
Sub-contractor	2	3.92%
Job title		
Project manager	16	31.37%
Construction manager	35	68.63%
Construction experience		
0–5 years	0	0%
5–20 years	45	88.24%
More than 20 years	6	11.76%
Company construction operations		
0–5 years	5	9.81%
5–20 years	28	54.90%
More than 20 years	18	35.29%

Table 8.5 Level of respondents' familiarity with exoskeletons in construction

S./no.	Device	Median	SD
1	Wearable robots	4.0	1.22
2	Active exoskeleton	4.0	1.24
3	Passive exoskeleton	4.0	1.23

8.4.2.1 Familiarity with Exoskeletons in the Construction

The result presented in Table 8.5 indicates that the pool of respondents is at least moderately familiar with exoskeletons (median = 4.0) and the devices' application in the construction domain. A closer look at responses to individual categories (active and passive) revealed no visible difference in the level of respondents' familiarity with the application of wearable robotics in construction. This level of familiarity further strengthens the reliability of critical insights obtained from the pool.

8.4.2.2 Can Exoskeletons Prevent WMSDs Per Body Region?

The result in Fig. 8.3 shows that, on average, respondents believe that wearable robotics could prevent 60% of construction-related WMSDs associated with all human body parts. This indicates that the respondents believe that exoskeletons are effective interventions, thereby making this class of technology a viable option in workers' personalized safety management. In addition, the spread highlighted in Fig. 8.3 reveals that shoulder and upper arm (90%), neck (78%), and hand (75%) are the top three regions wherein participants believe exoskeletons have the greatest potential to prevent WMSDs. Since most WMSDs affect the hands, wrists, elbows, neck, and shoulders [9], this result suggests that practitioners believe that exoskeletons could play a major role in preventing accidents impacting critical body regions.

8.4.2.3 Can Exoskeleton Prevent Accidents Associated with WMSD Risk Factors?

Results from the study indicate that using exoskeletons on construction projects could help prevent between 30 and 40% of accidents associated with critical risk factors (see Fig. 8.4). Specifically, practitioners selected improper lifting (40%), forcefulness and muscle effort (40%), repetitive work (39%), and taking improper working positions (39%) as top WMSD risk factors that could be prevented through increased application of exoskeletons. This relatively high accident prevention capacity supports the growing interest in the use of exoskeleton as an accident prevention tool. Therefore, the authors believe that researchers need to develop processes aimed at tackling inherent issues bordering usability, effectiveness, and return on investment.

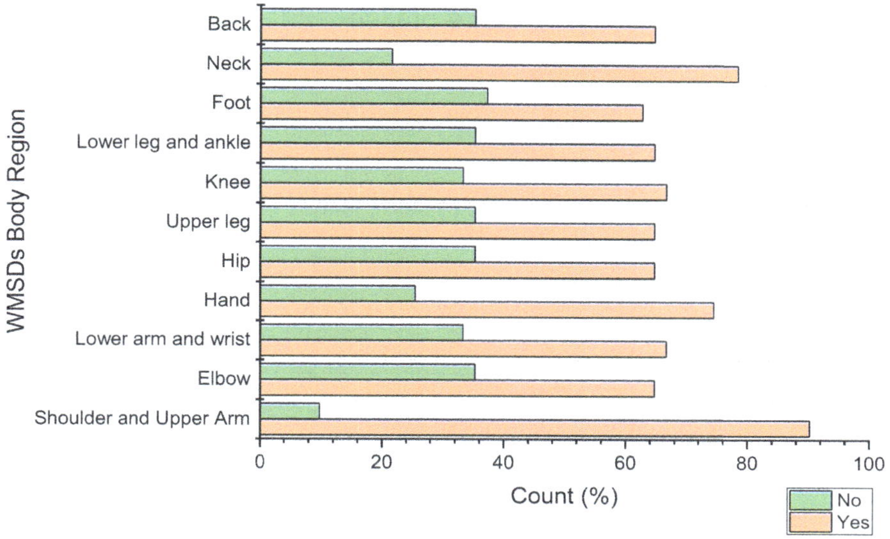

Fig. 8.3 Exoskeletons for preventing WMSDs per body regions

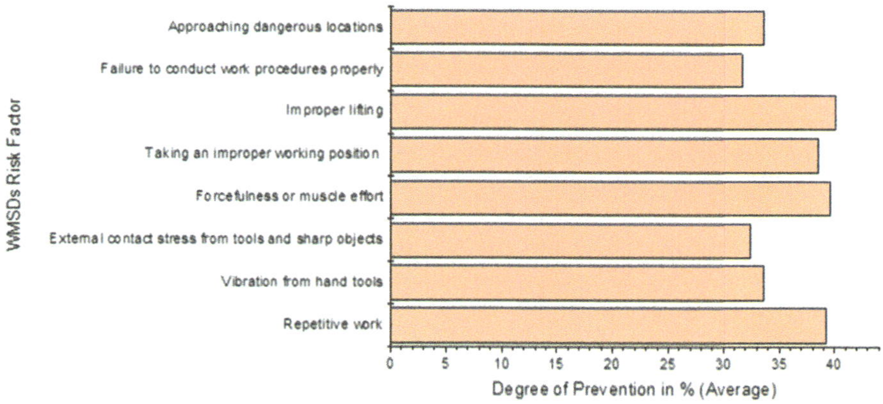

Fig. 8.4 Exoskeleton prevents accidents associated with WMSD risk factors

8.4.3 Usability of Exoskeleton for Flooring Tasks (Flex Lift)

Ten healthy male participants with no reported musculoskeletal issues or other related health that affects their ability to stand, walk, bend, and lift performed lab-simulated flooring task. Demographic statistics of participants as represented in mean and standard deviation were the following: ages 23 years ± 2, weight 147lb ± 23, and height 171cm ± 5. Body mass index of the participants ranged from normal to overweight 23kg/cm ± 3.

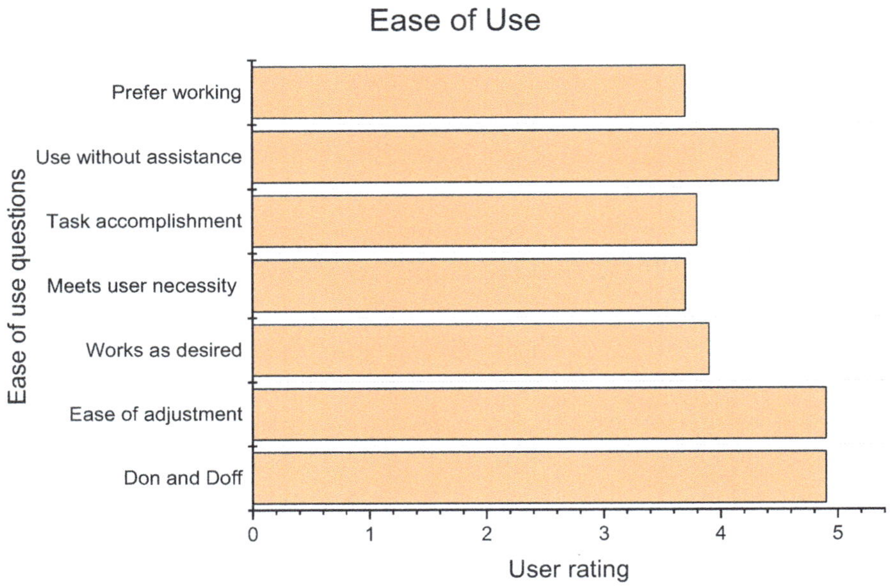

Fig. 8.5 Usability assessment: ease of use

Figures 8.5, 8.6, and 8.7 show the participants' ratings in response to the usability criteria (i.e., ease of use, ease of learning, and comfort) after performing the flooring task with the Flex Lift exoskeleton. Regarding ease of use, the participants rated don and doff and ease of adjustment very highly (4.9 each) alongside the ability to use the device without assistance (4.5). Importantly, the participants agreed (3.8) that the exoskeleton made their tasks easier to accomplish, which implies improved productivity. Other sub-elements attained a rating greater than 3.0, indicating a moderate level of agreement. Considering ease of learning, the participants agreed (greater than 4) with all sub-elements except having prior knowledge of the technology (1.7). This points to the need for extensive workers' training and widespread organizational awareness on the functionalities of exoskeletons.

Participants were asked three questions regarding their comfort with working with the Flex Lift exoskeleton for the flooring task. The questions relate to exoskeleton restriction in movement and interference with work and participants' satisfaction using the exoskeleton. The feedback from participants had a combined average rating of 2.2 ± 0.3. Participants disagreed that the exoskeleton restricts their movement and interferes with their work (1.5 each). However, participants indicated a satisfaction level of 3.5, which is above average. The minimal restriction to movement and interference to work could imply that the exoskeleton is suitable for tasks involving back bending and stooping postures inherent in flooring task. To enhance adoption of the Flex Lift exoskeleton in the construction industry, the authors would recommend optimizing the technology for comfort.

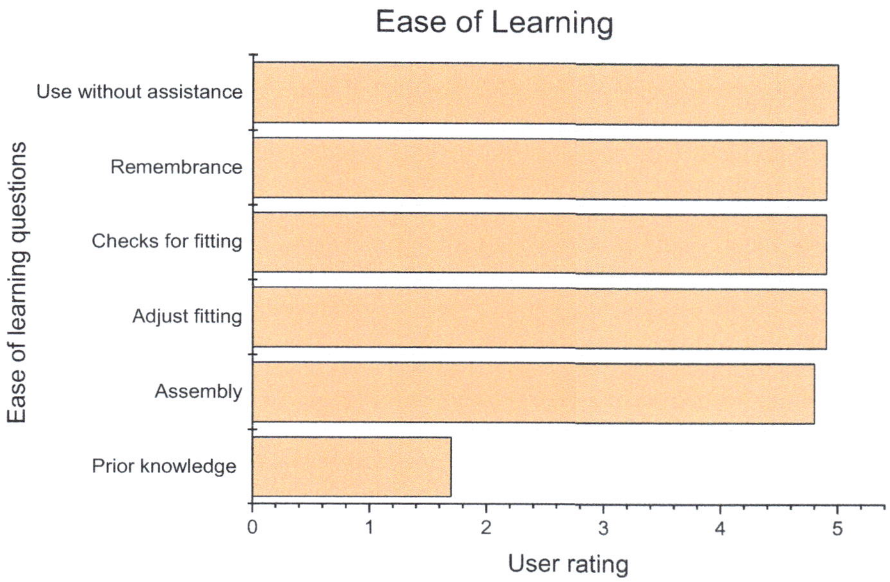

Fig. 8.6 Usability assessment: ease of learning

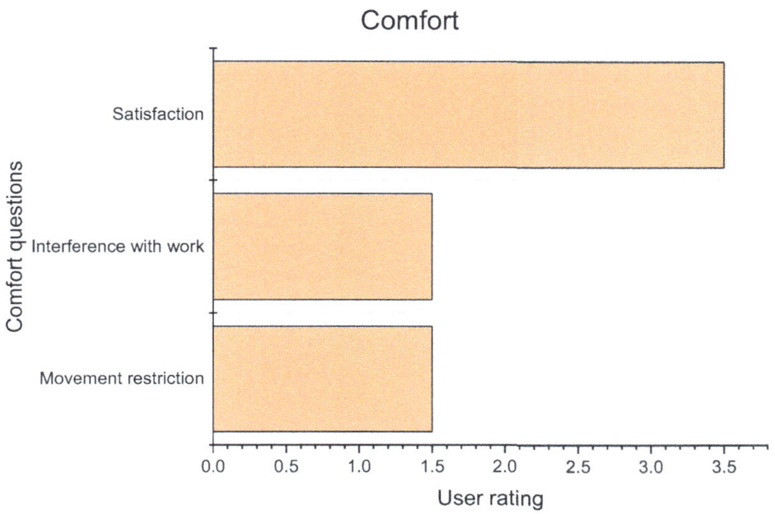

Fig. 8.7 Usability assessment: comfort

8.5 Conclusions

This paper presents the result of a multiphase study aimed at providing insight into the capability of commercially available wearable robots for preventing work-related musculoskeletal disorders (WMSDs). First, the authors presented 11 wearable robots with potential utility in construction operations; complete with applicable tasks, categories, and areas of application. Results from the study indicate that wearable robots could play a critical role in preventing injuries associated with lower and upper back, feet and ankle, wrist and hand, and shoulder. Regarding effectiveness of exoskeletons, project and construction managers surveyed that wearable robotics could prevent 60% of construction-related WMSDs associated with all human body parts and 30–40% of accident causes associated with WMSD risk factors. In addition, participants who used a wearable robot (Flex Lift) to perform a construction task (flooring) indicated that the exoskeleton was easy to use and easy to learn on how to use. Also, the exoskeleton posed minimal restrictions to participants' movement and mildly interfered with their tasks while improving their productivity. It is worthy to note that the results of the questionnaires only indicate the perception of the practitioners involved in this particular study. It does not guarantee that all wearable robots would be effective when used across different construction applications. Future studies should focus on performing controlled and live tests aimed at testing the effectiveness of exoskeletons in construction applications. In addition, it is important to assess the new safety risks introduced or enhanced by wearable robots when used to perform construction activities. In all, the present study provides critical insight needed to advance application of and research on wearable robots.

Acknowledgment This work was supported by the Pilot/Small Project Research Training Grant #T42OH008436 from the National Institute for Occupational Safety and Health (NIOSH), and the CPWR through NIOSH cooperative agreement OH009762. Its contents are solely the responsibility of the authors and do not necessarily represent the official views of CPWR or NIOSH. The grants are administered by CPWR, and the Deep South Center for Occupational Health and Safety at the UAB, School of Public Health.

References

1. S. Kim, A. Moore, D. Srinivasan, A. Akanmu, A. Barr, C. Harris-Adamson, D.M. Rempel, M.A. Nussbaum, Potential of exoskeleton technologies to enhance safety, health, and performance in construction: Industry perspectives and future research directions. IISE Trans Occup Ergonomic & Human Factor **7**(3-4), 185–191 (2019 Oct 2). https://doi.org/10.1080/24725838.2018.1561557
2. R. Kanan, O. Elhassan, R. Bensalem, An IoT-based autonomous system for workers' safety in construction sites with real-time alarming, monitoring, and positioning strategies. Automat Construct **88**, 73–86 (2018 Apr 1). https://doi.org/10.1016/j.autcon.2017.12.033

3. BLS (Bureau of Labor Statistics). National census of fatal occupational injuries in 2019. U.S. Dept. of Labor, Washington, DC. 2020. Retrieved from https://www.bls.gov/news.release/archives/osh_11042020.pdf on December 14, 2020.
4. P. Dasgupta, M. Sample, B. Buchholz, M. Brunette, Is worker involvement an ergonomic solution for construction intervention challenges: a systematic review. Theor. Issues Ergon. Sci. **18**(5), 433–441 (2017 Sep 3). https://doi.org/10.1080/1463922X.2016.1274452
5. B. Esmaeili, M.R. Hallowell, Diffusion of safety innovations in the construction industry. J Construct Eng & Manag **138**(8), 955–963 (2012 Aug 1). https://doi.org/10.1061/(ASCE)CO.1943-7862.0000499
6. I. Okpala, C. Nnaji, A.A. Karakhan, Utilizing emerging technologies for construction safety risk mitigation. Pract Period Struct Des Constr **25**(2), 04020002 (2020 May 1). https://doi.org/10.1061/(ASCE)SC.1943-5576.0000468
7. T. Bock, T. Linner, W. Ikeda, Exoskeleton and humanoid robotic technology in construction and built environment. Futur Humanoid Robot-Res & Appl, 111–144 (2012 Jan 20). https://doi.org/10.5772/27694
8. T. Bock, The future of construction automation: Technological disruption and the upcoming ubiquity of robotics. Automat Construct **59**, 113–121 (2015 Nov 1). https://doi.org/10.1016/j.autcon.2015.07.022
9. A. Karvouniari, G. Michalos, N. Dimitropoulos, S. Makris, An approach for exoskeleton integration in manufacturing lines using Virtual Reality techniques. Procedia CIRP. **78**, 103–108 (2018 Jan 1). https://doi.org/10.1016/j.procir.2018.08.315
10. C. Wang, L. Ikuma, J. Hondzinski, M. de Queiroz, Application of assistive wearable robotics to alleviate construction workforce shortage: Challenges and opportunities, in *Computing in Civil Engineering*, (2017), pp. 358–365. https://doi.org/10.1061/9780784480830.044
11. J.M. Delgado, L. Oyedele, A. Ajayi, L. Akanbi, O. Akinade, M. Bilal, H. Owolabi, Robotics and automated systems in construction: Understanding industry-specific challenges for adoption. J Build Eng **26**, 100868 (2019 Nov 1). https://doi.org/10.1016/j.jobe.2019.100868
12. Bureau of Labor Statistics. Occupational Safety and Health Definitions 2016. U.S. Dept. of Labor, Washington, DC. 2020. Retrieved from https://www.bls.gov/iif/oshdef.htm on December 14, 2020.
13. CPWR (Center for Construction Research and Training). The Construction Chart Book. 2016. Retrieved from https://www.cpwr.com/research/data-center/the-construction-chart-book/the-construction-chart-book-6th-edition-foreword/ on December 14, 2020.
14. BLS (Bureau of Labor Statistics). National census of fatal occupational injuries in 2019. U.S. Dept. of Labor, Washington, DC. 2020. Retrieved from https://www.bls.gov/iif/oshwc/case/msds.htm on March 1, 2021.
15. M. Middlesworth, The cost of musculoskeletal disorders (MSDs) [Infographic]. EgroSolution (2019)
16. M.D. Hossain, A. Aftab, M.H. Al Imam, I. Mahmud, I.A. Chowdhury, R.I. Kabir, M. Sarker, Prevalence of work related musculoskeletal disorders (WMSDs) and ergonomic risk assessment among readymade garment workers of Bangladesh: A cross sectional study. PLoS One **13**(7), e0200122 (2018 Jul 6). https://doi.org/10.1371/journal.pone.0200122
17. A.R. Anita, A. Yazdani, K.S. Hayati, M.Y. Adon, Association between awkward posture and musculoskeletal disorders (MSD) among assembly line workers in an automotive industry. Malays J Med & Health Sci **10**(1), 23–28 (2014)
18. CPWR (Center for Construction Research and Training). Preventing Musculoskeletal Disorders in Construction Workers. Retrieved from http://elcosh.org/document/1648/d000560/preventing-muskuloskeletal-disorders-in-construction-workers.html on December 14, 2020.
19. Wesslén J. Exoskeleton exploration: Research, development, and applicability of industrial exoskeletons in the automotive industry. 2018. http://urn.kb.se/resolve?urn=urn:nbn:se:hj:diva-40093
20. Kobayashi H, Uchimura A, Shiiba T. Development of muscle suit for upper body. In Proceedings 2003 IEEE/RSJ International Conference on Intelligent Robots and Systems

(IROS 2003) (Cat. No. 03CH37453) 2003 Oct 27 (Vol. 4, pp. 3624–3629). IEEE. https://doi.org/10.1109/IROS.2003.1249718
21. Cho YK, Kim K, Ma S, Ueda J. A robotic wearable exoskeleton for construction worker's safety and health. In ASCE Construction Research Congress 2018 (pp. 19–28). https://doi.org/10.1061/9780784481288.03.
22. S. Alabdulkarim, M.A. Nussbaum, Influences of different exoskeleton designs and tool mass on physical demands and performance in a simulated overhead drilling task. Appl. Ergon. **74**, 55–66 (2019 Jan 1). https://doi.org/10.1016/j.apergo.2018.08.004
23. J. Brooke, Sus: a "quick and dirty" usability. Usability Eval Indus **11**, 189 (1996 Jun)
24. S. Spada, L. Ghibaudo, S. Gilotta, L. Gastaldi, M.P. Cavatorta, Investigation into the applicability of a passive upper-limb exoskeleton in automotive industry. Procedia Manuf **11**, 1255–1262 (2017 Jan 1). https://doi.org/10.1016/j.promfg.2017.07.252
25. R. Bogue, Exoskeletons – a review of industrial applications. Indus Robot: Int J (2018 Aug 20). https://doi.org/10.1108/IR-05-2018-0109

Chapter 9
Worker-Aware Task Planning for Construction Robots: A Physiologically Based Communication Channel Interface

Yizhi Liu, Mahmoud Habibnezhad, and Houtan Jebelli

Abstract Unsafe workplace environment, labor shortage, and low productivity rate are among the most critical challenges faced by the construction industry. These ongoing challenges have prompted the construction research community to explore robotization as a new direction of improvement. Most of these endeavors have focused on human-robot collaboration (HRC) as the design and development of adaptable and scalable robots for the dynamic work environment of construction sites are excessively difficult. Although HRC can facilitate the design and implementation process, it can pose new threats to human workers' physical and mental health. One promising approach for a safe and productive HRC is that the co-workers' physiological responses are captured, interpreted, and infused as a new sensory information level into robotics optimization and planning systems. However, strict compliance of the robot to the system's prediction results may lead to unstable robotic adjustments because the robot modifies its performance by blindly following the imperfect system information. The present research seeks to optimize such reliance through physiologically aware signal classification algorithms that can appropriately establish harmony between worker physiology and robotic performance. The backbone of this algorithm is grounded in the fact that human physiological alterations are relatively gradual as opposed to electronic alterations of the systems that can happen in a fracture of a second. The proposed algorithm innovatively screens the machine-generated prediction results so as to produce logically sound decisions for robotic control and manipulations. To validate the algorithm in a real robotic system, the authors examined its performance in a collaboration between an unmanned terrestrial robot and a construction worker, with the total of eight subjects in the study. The terrestrial robot was designed to capture physiological signals from the workers' wearable biosensors, promptly translate them into high-level information, and modify its performance based on trained ML classifiers.

Y. Liu · M. Habibnezhad · H. Jebelli (✉)
Department of Architectural Engineering, Pennsylvania State University, State College, PA, USA
e-mail: ykl5505@psu.edu; mahmoud@psu.edu; hjebelli@psu.edu

H. Jebelli et al. (eds.), *Automation and Robotics in the Architecture, Engineering, and Construction Industry*, https://doi.org/10.1007/978-3-030-77163-8_9

The experiment results demonstrated the potentials of the proposed screening algorithm to improve the probability of the correct system decisions in adjusting the robot's performance. This research shows the algorithm can facilitate the stable implementation of robots within physiologically based HRC at construction sites. This algorithmic filtering process can also be extended to human-system interactions that can facilitate the design and development of interfaces between workers and computer systems.

Keywords Fault-tolerant mechanism · Human-robot collaboration · Physiological signals · Machine learning · Construction robots

9.1 Introduction

The stagnant productivity, workforce shortage, and a myriad of safety issues due to the dynamic characteristics of the construction sites have urged the research community to explore convergent solutions [1] with the capacity to transform construction from a traditional operation to an automized machinery-dominated process. As the construction industry gears toward industrialization, the adoption of single-task construction robots (STCRs) has become the mainstream of this automation effort. This robotization trend can be seen in several construction areas, such as constructing masonry walls, implementing rebars, and handling heavy materials. Supported by the advent of artificial intelligence, robotics hardware and software, and sensitive peripheral robotic sensors, the industry has witnessed the manifestations of semi-automated mason (SAM) executing bricklaying tasks with five times the manpower [2], rebar-tying robot (TyBot) performing the steel reinforcement tasks with 50% less workforce [3], and Material Unit Lift Enhancer (MULE) lifting construction materials weighing up to 150 lbs. [4].

Although the proposed robotic solution appears poised to ameliorate the ever-existing safety and productivity challenges in the construction industry, workers' interactions and supervision are fundamental requirements for successful implementations of these pre-coded machines in the human-populated environments of construction sites [5]. This is because these single-task robots require multi-faceted design consideration pertinent to a comprehensive perception of their dynamic surrounding, the ongoing task, and the resources necessary to complete the operation. Of course, these requirements can be drastically challenging and demands human dexterities and judgment.

Chiefly, the interactions between human workers and robots can trigger new safety concerns, specifically when robots and workers collaborate to complete a task in the dynamic environments of construction sites. There is always a chance of collision with robotic parts when workers are jointly operating close to collaborative robots [5, 6]. In addition, the physical and mental abilities of workers may not match those of collaborative robots, which can lead to workers' physical and mental fatigue or excessive stress [7]. To overcome these barriers, seamless and reliable communications need to be established between the two peers, allowing the robot to intelligently perceive and respond to the human co-worker's physical and mental states [8–10].

Emerging sensing approaches have opened doors to establishing reliable and intuitive communication channels between collaborative robots and the human workforce. As investigated in recent studies [8, 11–14], intelligent cameras [11], tactile and force-torque sensors [12], and the laser rangefinder [13] have empowered robots with the ability to capture and process explicit information of the worker, such as position, contact force, and moving direction. These sensory data will be infused into the robotic control system to execute appropriate actions or generate informative feedback to the worker. Moreover, the emergence of advanced biosensors – wearable electroencephalogram (EEG), electromyography (EMG), heart rate, and electrodermal (EDA) sensors – has set the stage for constructing implicit communications between robots and workers with the capacity to capture workers' physiological responses, evaluate their mental and physical conditions, and proactively execute safety interventions [8, 14, 18].

Among these human-robot communication strategies, the investigations of employing wearable biosensors to develop physiological signal-based HRC paradigms are gaining increasing attention. The representative application of this HRC method is to provide the robot with the anthropomorphic perception system, thereby forming an intelligent partnership between humans and robots. These physiologically aware approaches allow robots to automatically adjust their performance level to match workers' physiological status (e.g., emotion, cognitive load, and physical fatigue). They promise a safe and reliable collaboration between the two counterparts [17, 19], providing the necessary support for workers' health and well-being during HRC [20, 21].

Significantly, due to the noise in the biosignals and their time-variant characteristic (e.g., non-stationary issues of EEG and EMG signals [22, 23]), the built-in machine learning (ML) classifiers – the core of the physiologically based HRC paradigms – are incapable of always providing flawless decoding performance to the robotic operating system [20, 22, 23]. As such, the classification errors of these predictive classifiers will lead to incorrect and unpredictable robotic modulations. These incorrect classifications will be reflected as fluctuations in the robotic movement, which can severely influence the robots' mechanical performance in long-term use. Furthermore, the undesirable robotic adjustment will threaten workers' safety, especially in the proximity of collaborative robots. For example, erroneous adjustments will modify the actions (configuration and trajectory) of robots in a manner that is contrary to the intention of workers, which will cause severe injuries [24], like the pinch injuries (part of the worker's body is pinched between robots and other equipment) [11] and impact injuries (the collision between the workers and robots) [24, 25]. Besides, the erroneous adjustments will prevent workers from considering their collaborative robots as reliable humanoid partners. The unreliable partnership will impose extremely high working pressure on workers and adversely affect their mental health [26, 27].

To address these limitations, the authors propose an innovative fault-tolerant algorithm capable of screening the undesirable (incorrect) robotic adjustment triggered due to decoding errors of ML classifiers. The backbone of the proposed mechanism is grounded in the fact that human physiological states remain invariant in a relatively short period [28, 29]. To examine the feasibility of the proposed approach, the authors compared the performance of a physiologically based robotic modulation system with and without the proposed fault-tolerant algorithm. To that end, eight subjects were recruited to conduct a collaborative masonry task with an unmanned ground vehicle (UGV). During the collaboration task, the subjects' physiological signals were continuously obtained using wearable EEG and EMG biosensors. Subsequently, the recorded signals were interpreted into distinct physiological conditions using an artificial neural network (ANN) classifier. The classification results were then utilized by the UGV's robotic systems as follows: (1) the speed of the UGV was modulated by the intact prediction results, and (2) the speed of the UGV was regulated based on the prediction results filtered by the proposed algorithm. The comparison results of the two scenarios revealed that the algorithm could significantly improve the confidence level of the prediction outcomes, thus enhancing the stability and performance of the UGV during human-robot collaborations. The findings demonstrated that the proposed algorithm realizes the functionality of human physiology and uses this knowledge to hone the physiologically based robotic adjustment. This knowledge-based adjustment holds much promise in establishing a robust collaboration between workers and robots, paving the way for safe, efficient, and reliable collaboration at dynamic and unstructured construction sites.

9.2 Physiologically Based Human-Robot Collaboration (HRC)

9.2.1 Current Investigations on Physiologically Based HRC

In recent years, there has been an increasing amount of literature on employing physiological signals to establish the communication channels between human subjects and robots, both in controlled environments of laboratories as well as naturalistic working environments. According to these investigations, the use of human brainwaves (EEG) as communicative pipelines between humans and robots has the potential to boost the efficiency and reliability of human-robot interactions. For instance, Salazar-Gomez et al. stated that a corrective mechanism based on the error-related potential (ErrP) feedback measured from human EEG signals has high potentials to correct the motion trajectories of the Rethink Robotics Baxter [14]. Similarly, Liu et al. established a brainwave-driven HRC framework capable of interpreting workers' EEG signals into various levels of psychophysiological states, whereby the performance of the robot can be regulated to match the worker's cognitive load [16].

In addition to EEG signals, a handful of approaches have focused on leveraging EMG signals to achieve reliable and safe HRC strategies. Bu et al. conducted several human-robot experiments and showed that incorporating EMG signals as communicative pathways between humans and robots can help robots better understand subjects' motions and hand positions during collaborative tasks [8]. The results of their study unfold the potential role of EMG signals in enhancing human-robot collaboration by avoiding collisions with the robotic parts. Likewise, as highlighted by Peternel et al., the integration of fatigue prediction models with the KUKA robot's control system empowers the robot with the ability to evaluate workers' physical fatigue from their EEG signals and accordingly modify its working pace [30]. Furthermore, a multitude of other studies proposed several HRC frameworks driven by ECD, EDA, and heart rate physiological signals [17, 31, 32] to facilitate the communication between humans and robots and consequently establish more reliable and seamless human-robot collaborations. The underlying idea behind these investigations is that workers' implicit information obtained from their physiological signals can provide meaningful cues regarding their mental and physical conditions, which can inform the collaborative robot and adjust the robotic performance.

9.2.2 *Representative Structure of the Physiologically Based HRC*

The majority of the physiologically based HRC investigations primarily follow a similar paradigm (Fig. 9.1), interpreting biosignals into meaningful cues to which the robot can accordingly respond. More precisely, most of these approaches consist of two core components: (1) a translating engine capable of decoding biosignals into high-level information by denoising [14, 33, 34], processing [16, 35], and classifying [9, 16, 36] the signals (Fig. 9.1a) and (2) a transformer engine capable of converting the classification results into robotic adjustment commands using a robotic operating system and mechatronics principals (Fig. 9.1b). The integration of these two components enables the robot to modify its performance to match human capacity. For robotic arms, the adjustment will be based on the modulation of the joint angles employing built-in inverse and forward kinetic mechanisms [14, 37, 38]. As for vehicle robots, the adjustment process is according to the modification of the movement speed and direction parameters of the robot kinematic model, which includes motion equations of omnidirectional and differentially steered vehicle robots [16, 39, 40]. These adjustments aim to establish a harmonized, physiologically aware collaboration between humans and their mechanical counterparts [9, 13, 14].

Notably, the ML classifier plays a critical role in these approaches, classifying the human collaborators' physiological signals and generating the classification results that will be converted into the robotic adjustment commands (green box in Fig. 9.1). In other words, the classification accuracy of the ML classifier will

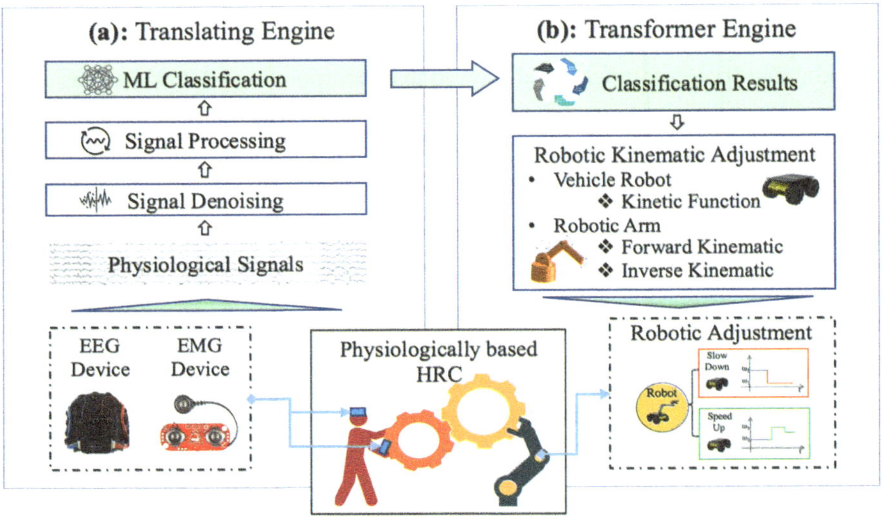

Fig. 9.1 Typical structure of the physiologically based human-robot collaboration systems

exclusively determine the adjustment accuracy of the collaborative robots in the physiologically based HRC system [9, 14, 36, 41]. Table 9.1 summarizes some of these classifiers along with their decoding performance.

9.3 Fault-Tolerant Algorithm in Physiologically Based HRC

As demonstrated in Table 9.1, the decoding accuracy of the built-in ML classifiers in physiologically based HRC approaches does not transcend beyond 92.9%. Moreover, the non-stationary issues associated with physiological data can further reduce the reported prediction accuracies. Therefore, these techniques will produce undesirable robotic adjustments. The strict reliance on the prediction results will produce unexpected robotic movements and adversely affect safety and trust between humans and workers [45, 46]. To address this limitation, the authors propose a fault-tolerant filtering algorithm that receives a stream of prediction classifications from the ML model, modulates the data, and generates "logically" sound predictions. The theoretical basis of the proposed fault-tolerant algorithm is derived from the "instantaneous load over time" diagram [29] in cognitive load theory (CLT) and ratings of perceived exertions (RPE) in physical fatigue measurement [47], stating that the physiological states of humans remain invariable in a short period (second-based). In other words, prompt variations in the classification results are most likely triggered due to prediction errors and not the variations in the physiological signals [28, 29]. With this prior knowledge, the schematic diagram of the fault-tolerant mechanism is shown in Fig. 9.2.

Table 9.1 Representative examples of physiological signal-based HRC studies

Physiological signal-based HRC	Classifier	Decoding accuracy	References
Decodes subjects' EMG signals to allow the robot to adjust its trajectories	LLGMN*, Hybrid BN*, and MP-NN*	85.1%; 92.9%	[9]
Regulates the joint angle of the robotic arm based on the decoded results of subjects' EMG signals	ANN*	87%	[42]
Empowers collaborative robots to plan the sequence of car engine assembly task based on the recognition of workers' EEG signals	ANN	81.4%	[41]
Enables robots to self-correct their motion mistakes based on the subjects' EEG signals	EN*	86.51%	[14]
Manipulates the trajectories of the robotic arm through subjects' EEG signals	LDA*	49.45%	[43]
Enables robots to understand subjects' engagement state through Heart Rate signals	KNN*	84.73%	[44]
Analyzes EDA* signals to make robots understand subjects' anger and happiness	LR*-SVM*	77.5%	[31]
Uses EDA and ECG* signals to enable robots to perceive their human teammates' cognitive load	SVM; RF*; NB*	92.9% 78.6% 57.1%	[32]
Adjusts the behavior of the robotic arm according to workers' fatigue level through ECG signals	Fuzzy classifier	74.4%	[17]

*Abbreviation: *LLGMN* log-linearized Gaussian mixture network, *BN* Bayesian network, *MP-NN* motion prediction neural network, *ANN* artificial neural network, *EN* elastic network, *LDA* linear discrimination analysis, *KNN* K-nearest neighbors, *EDA* electrodermal, *LR* logistic regression, *SVM* support vector machine, *ECG* electrocardiography, *RF* random forest, *NB* naïve Bayers

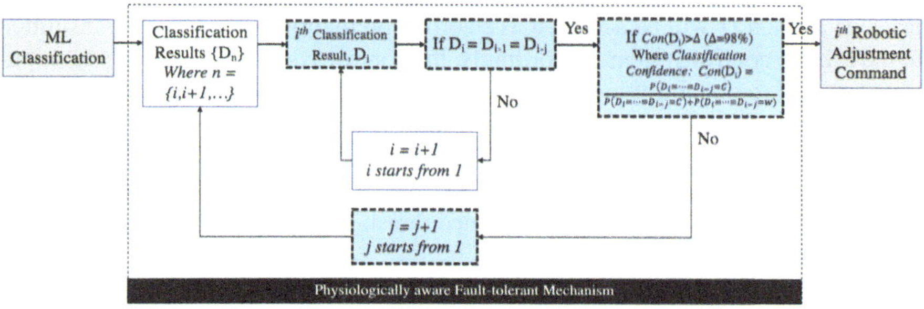

Fig. 9.2 Fault-tolerant mechanism for physiologically based HRC

The main elements of the proposed fault-tolerant mechanism (dash line and blue box in Fig. 9.2) are as follows: (1) the receiver module obtaining the classification outcomes; (2) the comparison module comparing consecutive classification results; (3) the confidence evaluation module of classification results (classification confidence), $Con(D_i)$, assessing the number of consecutive points required to meet the predetermined confidence level; and (4) the fault tolerance level j (starts from 1), determining the number of consecutive data points (classification results) that need to be considered in the comparison module. The receiver module obtains the classification results, D_i ($i = \{1, 2, \ldots, n\}$), from the predictive models. This component indexes the array of classification results and prepares the data for comparative analysis. In the comparison module, the ith classification results of workers' physiological signals, D_i, will be compared to its j previous consecutive data, $\{D_{i-1}, \ldots, D_{i-j}\}$. If these $j + 1$ predecessors are equal, then the classification confidence of D_i will be evaluated in the confidence evaluation module using equation 1. Otherwise, the mechanism will increment i by one unit until it finds $j + 1$ identical consecutive results:

$$\textit{Classification Confidence}: Con\left(D_i\right) \quad \frac{P\left(D_i = D_{i+1} = \ldots = D_{i+j} = C\right)}{P\left(D_i = D_{i+1} = \ldots = D_{i+j} = C\right) + P\left(D_i = D_{i+1} = \ldots = D_{i+j} = W\right)} \tag{9.1}$$

In equation 1, C indicates the correct classification results in line with the actual physiological states of the worker, whereas W shows otherwise. Probability P $(D_i = D_{i+1} = \ldots = D_{i+j} = C)$ and P $(D_i = D_{i+1} = \ldots = D_{i+j} = W)$ can be then calculated as the product of the probability of the correct and wrong classification from D_i to D_{i+j}, respectively. For each classification result, D_i, the probability of correct classification, $P(D_i = C)$, is equivalent to the classification accuracy of the ML classifier; the probability of misclassification, $P(D_i = W)$, is then 100% – classification accuracy of the classifier. Equation 9.1 calculates the conditional probability that D_i is correct when it is equal to its j predecessors.

After calculating the value of $Con(D_i)$, it will be compared with a user-defined threshold, Δ ($\Delta = 98\%$ in this study). If the classification confidence $Con(D_i)$ exceeds Δ, then D_i will be accepted as a correct prediction outcome and transmitted to the robotic control system. Accordingly, the control system will properly manipulate the robot (e.g., modifying the configuration, working pace, and trajectory) with a confidence level higher than Δ. Conversely, if $Con(D_i)$ does not surpass Δ, the fault tolerance level will be incremented by one unit, $j = j + 1$, according to which $Con(D_i)$ will be re-evaluated. In this regard, the final fault-tolerant setting can be determined based on a specific user-defined threshold Δ and the decoding accuracy of the ML classifier.

For example, for an ML classifier with 80% prediction accuracy, if Δ is set to 98% confidence level and two consecutive prediction results (j starts from 1) are considered for the comparison, the result of $Con(D_i)$ will be lower than Δ, which is $Con\left(D_i\right) = \frac{0.8^2}{0.8^2 + 0.2^2} = 94.1\% < \Delta$. Afterward, the algorithm will increase the fault tolerance level to two (j starts from 2), indicating that the comparison module will use three consecutive prediction results. At this level, if three successive prediction results belong to the same prediction class, the value of $Con(D_i)$ function will be $\frac{0.8^3}{0.8^3 + 0.2^3} = 98.5\%$, which will be greater than Δ.

Based on the proposed physiologically aware algorithm, the robotic performance will be regulated with a high confidence level, low variations in the prediction results, and smooth transitions between various performance levels. This filtering algorithm can profoundly alleviate variations in robotic adjustments due to misclassifications of predictive models and consequently reduce undesirable robotic adjustments.

9.4 Case Study to Evaluate the Performance of the Proposed Fault-Tolerant Mechanism

To examine the feasibility of the proposed fault-tolerant mechanism in harmonizing the collaboration between the workers and robots, an experiment was designed with two similar collaborative tasks, EEG- and EMG-based HRC tasks. Eight able-bodied subjects (age: mean = 24.3, SD = 1.66) were recruited to perform the two collaborative tasks. In the first task, subjects were equipped with a wearable EEG sensor (32-channel Emotiv Flex [16]), instructed to work with an unmanned ground vehicle (UGV) jointly, and constructed a masonry wall (5-by-6 blocks). In the second task, subjects were equipped with wearable EMG sensors (muscleBAN [48]) and asked to build the same wall in collaboration with the UGV. Fig. 9.3 demonstrates the wearable sensors and the layouts of the two tasks.

The EEG-based HRC task consisted of 30 rounds of collaboration with three low, medium, and high cognitive load levels (10 rounds for each level), imposed on subjects using an N-digit task. In each round, subjects were required to stand on the working area, memorize an N-digit number (N can take the values of 1, 3, and 8 for low, medium, and high cognitive loads, respectively [49–51]), and then pick up a concrete block from the container of the UGV and appropriately lay the block to construct the wall. The UGV was designed to automatically return to the loading area, where it would be loaded with another concrete block. At the end of each round, subjects were prompted to type the memorized N-digit number on the laptop (Fig. 9.3-Task A).

During the EMG-based HRC task, subjects were required to perform 30 rounds of masonry task, 15 of which were performed under low physical demands and 15 under high physical demands. To simulate low physical demanding activity, subjects were asked to pick up relatively light-weight construction materials (Fig. 9-Task B3: 5-lbs. concrete block) from the UGV's container. Similar to the previous task, subjects would then lay the block to build the wall. Meanwhile, the UGV would return to the loading area to be loaded with the same-weight concrete block. To simulate high physical demanding activity, subjects were required to pick up and lay a 34-lbs. concrete block (Fig. 9.3-Task B) to construct a 3-by-5 block wall.

The trajectories of the UGV robot were programmed and controlled by robotic operating system (ROS). More information about the ROS structure can be found in [16, 52]. During the collaboration tasks, the movement speed of the UGV was determined based on the classification results of the ML classifier (ANN classifier in this study) for subjects' physiological signals. If the robot strictly complies with the ML

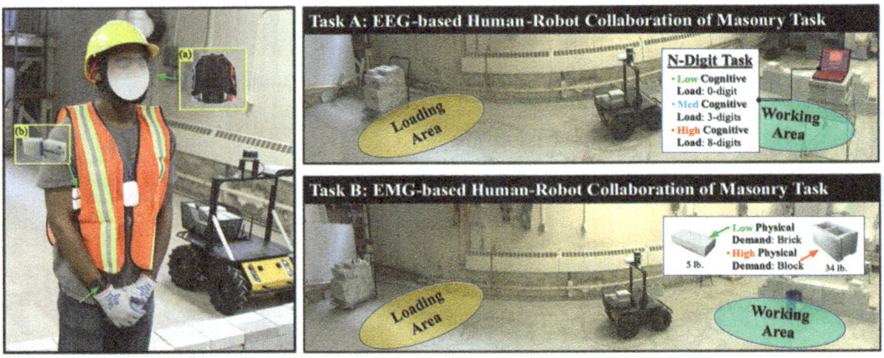

Fig. 9.3 EEG and EMG wearable biosensors and layouts of the designed 30-rounds EEG- and EMG-based HRC tasks

classifier results, then the movement speeds will be as follows for the EEG- and EMG-based HRC tasks:

(1) EEG-based HRC task: for low cognitive load, the UGV robot will increase its speed and assist its human partner in completing the round in the 20s (3 rounds/min) to reduce subjects' spare time; for high cognitive loads, the UGV robot will slow down its pace and enable subjects to finish this round in 40 seconds (1.66 rounds/min), so that providing subjects more time to recovery. As for the medium cognitive load, the UGV will maintain its initial working speed, and the duration of each round is the 30s (2 rounds/min).
(2) EMG-based HRC task: during the low physical demand rounds of the HRC task, the working speed of UGV robots was set to 20s per round (3 rounds/min). At the high physical demand level, the robot will self-adjust its speed to 40 seconds per round (1.5 rounds/min), enabling subjects to recover from their physically demanding task.

Both experimental tasks were conducted using two different systems, one that directly manipulates robotic kinematics following the translated biosignals (HRC_1 system) and one that performs the robotic adjustment using the proposed fault-tolerant algorithm, with the user-defined threshold of Δ equal to 98%, embedded in the robotic control system (HRC_2 system). The structures of these two HRC mechanisms are shown in Fig. 9.4. As can be seen in the figure, by employing the HRC_1 system, the workers' physiological signals (EEG, EMG) will be decoded by the ML classifier (ANN) and directly translated into the robotic command for adjusting the performance of the UGV (Fig. 9.4a). However, using the HRC_2 system, the robot's movement speed will be merely changed if the confidence level of the system for correctly interpreting the biosignals is above 98% (Fig. 9.4b). More details regarding the adjustment process for the two tasks are presented in Sect. 9.5.

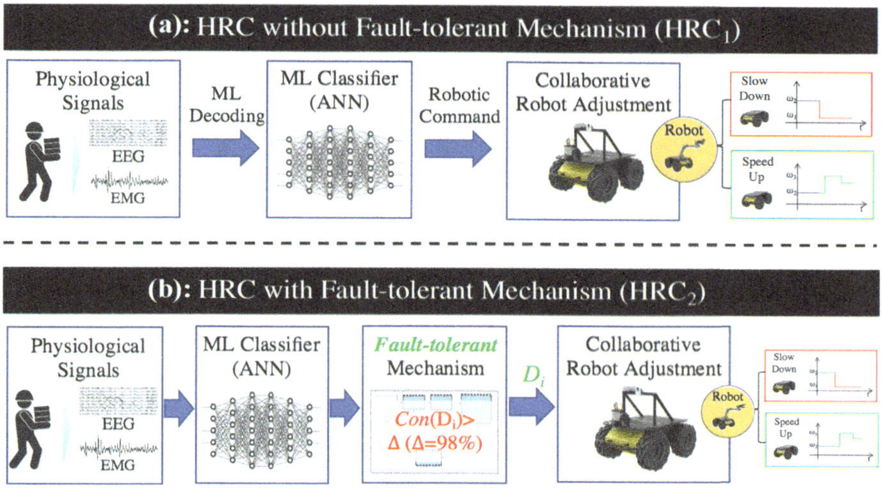

Fig. 9.4 (**a**) Structure of the physiologically based HRC without the fault-tolerant mechanism used in the designed HRC tasks (**b**) Structure of the physiologically based HRC with fault-tolerant mechanism used in the designed HRC tasks

9.5 Results and Discussion

9.5.1 Results of the EEG-Based HRC Masonry Task

After all the physiological data were collected from the subjects, an in-house Python code was developed to read and tabulate the information into structured data. These data contain information about the prediction results of the ANN classifiers as well as the kinematic information of the robot during the experimental procedures. Fig. 9.5a illustrates the ANN classifier's recognition capability in distinguishing various levels of cognitive load with 83.71% prediction accuracy. Moreover, it visualizes the decision boundary of the ANN classifier, which reflects the capability of the well-calibrated ANN classifier to efficiently decode the nonlinearly distributed physiological signals across subjects.

By employing the well-trained ANN classifier, the physiologically based HRC mechanism in both HRC_1 and HRC_2 systems enabled the collaborative UGV robot to capture subjects' EEG signals, interpret them into distinct cognitive load levels, and accordingly adjust the operating speed of the robot. Yet, the robotic modulation was different across the two systems. For each subject (eight in total), Table 9.2 reports the speed adjustment accuracy of the UGV robot during the collaborative block-laying task using the HRC_1 and HRC_2 systems. By averaging the adjustment accuracy of all eight subjects, it can be shown that the fault-tolerant algorithm can promote the confidence level of the prediction outcome by 23.5%, a significant jump from HRC_1 (74.59%) to HRC_2 (92.09%).

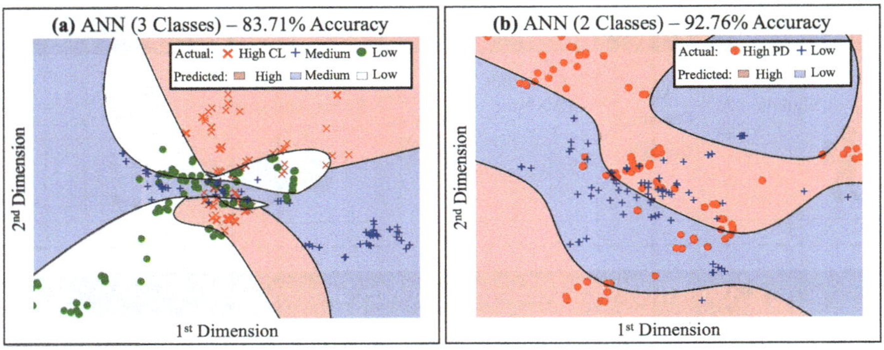

Fig. 9.5 Decoding accuracy of the ML classifier (ANN) in (**a**) EEG-based HRC task and its decision boundary and in (**b**) EMG-based HRC task and its decision boundary

Table 9.2 The adjustment accuracy of the UGV robot working pace in the EEG-based HRC masonry task for each subject

Subject no.	Accuracy by employing HRC_1	Accuracy by employing HRC_2	Subject no.	Accuracy by employing HRC_1	Accuracy by employing HRC_2
1	73.3%	90.0%	5	73.3%	90.0%
2	76.7%	93.3%	6	80.0%	93.3%
3	76.7%	90.0%	7	76.7%	96.7%
4	70.0%	96.7%	8	70.0%	86.7%

Figure 9.6 shows an example of speed adjustment of the UGV robot during the collaborative block-laying task using the HRC_1 and HRC_2 mechanisms for one randomly selected subject. The blue line indicates the actual speed adjustment of the robot triggered by the system. In contrast, the red dash line represents the desirable (expected) robotic adjustment corresponding to the stimulated cognitive load using the N-digit task. As shown, during all the 30 rounds of the task, the HRC_2 system performed substantially better than the HRC_1 system. More specifically, the collaborative system integrated with our proposed algorithm resulted in 93.3% accuracy of following the desirable robotic adjustment compared to the 76.7% accuracy of the HRC_1 system. Fig. 9.6a presents the adjusted working speed of the collaborative robot by leveraging the HRC_1 system. The comparison between the adjustment performance of the two systems shows that the proposed fault-tolerant algorithm can filter out five undesirable robotic adjustments, enhancing the accuracy of robot adjustment by 27% ($\frac{93.3-76.7\%}{76.7\%}$) in the EEG-based HRC task.

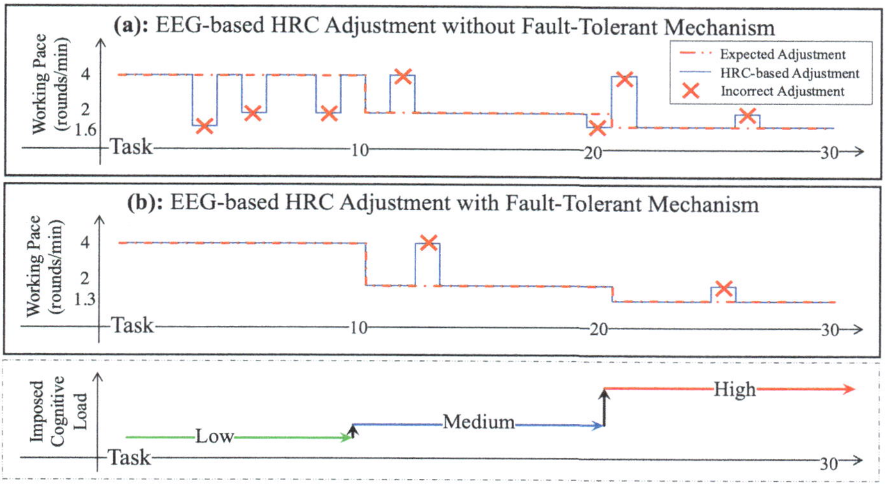

Fig. 9.6 UGV's speed adjustments results for 30 rounds of EEG-based HRC task for one randomly selected subject using (**a**) direct manipulation system and (**b**) the fault-tolerant mechanism

9.5.2 Results of the EMG-Based HRC Masonry Task

As explained above, all the experimental data were recorded, tabulated, and analyzed, specifically the information about the ANN classifier's prediction accuracy and the resulting performance of the UGV in coping with various levels of workers' mental and physical workload demands. Fig. 9.5b illustrates the ANN classifier's decoding performance, fused with both robotic control systems, in classifying the low and high physically demanding block-laying tasks based on the subjects' retrieved EEG signals. As can be seen in the figure, the decision boundary delineates the efficacy of the ANN classifier in binarily classifying the EMG signals with 92.76% accuracy. Even though the model's prediction accuracy holds promise to establish strong communication between humans and robots, the misclassifications of the model can lead to robotic adjustments that are not in line with the correct physical demands of the task.

To compare the adjustment performance of the two HRC system (HRC_1 and HRC_2) during the EMG-based block-laying task, the movement speed of the UGV was recorded for both systems. As shown in Fig. 9.7a, for one randomly selected subject, without applying the proposed fault-tolerant mechanism to the physiologically based HRC system, in all 30 rounds of the EMG-based HRC task, HRC_1 will cause six undesirable speed adjustments of the UGV robot. On the contrary, by employing the fault-tolerant mechanism, HRC_2 will only generate one adjustment errors in the same EMG-based task (see Fig. 9.7b). Similar to Sect. 9.5.1, for all eight subjects, the authors have reported the comparison of the robot adjustment accuracy using the HRC_1 and HRC_2 systems during the EMG-based task. As reported in Table 9.3, the HRC system without the fault-tolerant algorithm (HRC_1) can only allow the collaborative UGV

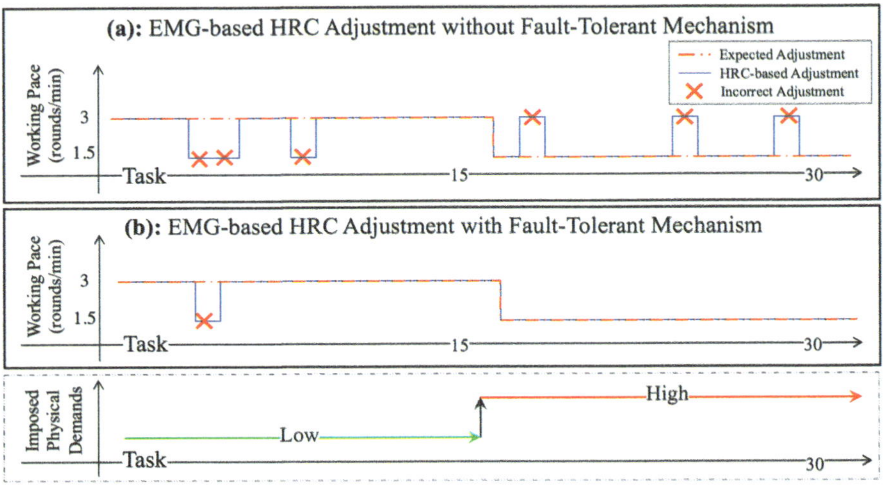

Fig. 9.7 UGV's speed adjustments result for 30 rounds of EMG-based HRC task using (**a**) direct manipulation system and (**b**) the fault-tolerant mechanism

Table 9.3 The adjustment accuracy of the UGV robot working pace in the EMG-based HRC masonry task for each subject

Subject no.	Accuracy by employing HRC_1	Accuracy by employing HRC_2	Subject no.	Accuracy by employing HRC_1	Accuracy by employing HRC_2
1	83.3%	93.3%	5	76.7%	96.7%
2	80.0%	96.7%	6	80.0%	96.7%
3	86.7%	90.0%	7	76.7%	93.3%
4	76.7%	96.7%	8	73.3%	90.0%

robot to adjust its working pace with an average accuracy of 79.18%. However, the HRC system augmented with the fault-tolerant algorithm (HRC_2) enhances the average adjustment accuracy of the robot to 94.18% for all subjects. Therefore, compared with the robotic adjustment performance triggered by HRC_1, the average adjustment accuracy of the UGV driven by HRC_2 is elevated by 18.9%. Along with the promising results reported in Sect. 9.5.1, such accuracy enhancements reveal the feasibility of the proposed fault-tolerant mechanism in establishing the harmonized, robust, and safe HRC at the construction sites.

9.5.3 Discussion on the Performance of the Fault-Tolerant Algorithm

As we move forward with adopting collaborative robots at construction sites, the prominent role of perceptive robotic control systems in continuously understanding workers' mental and physical states and appropriately adjusting the robot's performance becomes apparent. The findings of the present study suggest that the integration of the proposed fault-tolerant algorithm with these physiologically aware perceptive robotic systems can substantially augment the performance of the system in regulating the robotic behavior. Our findings revealed that the proposed algorithm could harmonize the behavior of adaptive robots by detecting the correct patterns of variations in human physiological conditions and allowing the robotic control system to act upon these predictions. The experiment results advocate for such a performance, demonstrating that the fault-tolerant algorithm produces robust control system decisions. As such, the algorithm puts more weight on consistent performance rather than prompt manipulations but with irrational robotic behaviors (movement speeds for this study).

The algorithm provides a confidence level for accepting the prediction accuracies of the data-driven models in the system. This remark by no means suggests that each specific misclassification of the classifiers can be recognized. But it articulates that the logical sequence of classification outcomes can provide a high level of confidence in accepting the results and manipulating the robot's working pace. In other words, the prediction accuracy of the predictive models cannot be improved for each classification decision, but the underlying patterns of the classifications can elucidate whether there is a consistency between the prediction results and the rate of physiological changes.

Despite promising results, the proposed algorithm is bound to a notable limitation, the potential time-lags potentially arising during highly deterministic decisions of the fault-tolerant algorithm (i.e., $\Delta \geq 98\%$). As demonstrated in Fig. 9.8a, during the EEG-based HRC task, the HRC_2 required an average duration of 5.15s to modify the speed of the UGV. This duration is about four times longer than that obtained from the HRC_1 system for the same task (1.16s). Similar observations can be made for the EMG-based HRC task in which the average duration for generating commands is around three times more than that without the proposed algorithm (Fig. 9.8b).

The delay is mainly due to the imbalance between the original decoding accuracy of the ML classifier and the user-defined threshold (Δ). For example, if the decoding accuracy of the ML classifier is 60%, the algorithm needs to evaluate ten consecutive prediction results to achieve an over 98% certainty ($\Delta = \frac{0.6^{10}}{0.6^{10}+0.4^{10}} > 98\%$), whereas if the decoding accuracy of the ML classifier is 80%, it requires to consider three consecutive data points ($\Delta = \frac{0.8^{5}}{0.8^{5}+0.2^{5}} > 98\%$). Therefore, the physiologically based HRC system with a lower decoding accuracy (determined by ML classifier) needs to evaluate more consecutive decoded results to generate a robotic adjustment command. The increase in these consecutive points can lead to high computational complexity, significantly worsening computational memory usage and notably extending the computational time [53, 54].

One plausible approach to mitigating these potential time-lags is optimally selecting the confidence level in the proposed fault-tolerant algorithm. This selection should be made based on task specificity and available resources for

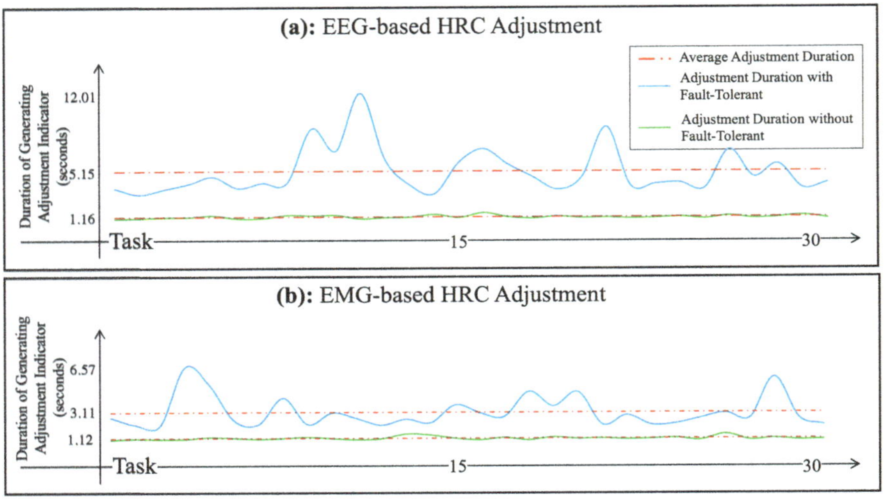

Fig. 9.8 Comparison results of the duration of generating robot adjustment indicators with and without applying proposed fault-tolerant mechanism in physiologically based HRC, for one randomly selected subject. (**a**) comparison results in the EEG-based HRC task. (**b**) Comparison results in the EMG-based HRC task

computation power and data acquisition rates of the sensors, as well as the adjustment rate necessary for effectively harmonizing collaboration between humans and robots.

9.6 Conclusions

This study sets out to propose and verify a fault-tolerant algorithm for improving the stability and reliability of the physiologically based HRC systems. Grounded in the knowledge of human physiology, the proposed algorithm was able to continuously obtain a set of consecutive classification results from the system's predictive classifiers, promptly compare these together, and subsequently decide whether or not the classification results should be passed to the robotic control system for proper robotic modulation. The findings showed that this simple yet effective algorithm could filter out "irrational" classifications and better match collaborative robots' performance with their human partners' physiological states (cognitive load and physical demands). Based on the experimental results, the integration of the proposed fault-tolerant algorithm with a physiologically based HRC system reduced undesirable robotic manipulation and improved the consistency of the robotic adjustments by approximately 20%.

The primary concept of this algorithm is that the classification results of physiological signals should follow a rational pattern in line with the capacity of the human body in rapidly altering its physiological states. This study exploits this biological limitation to detect physiologically acceptable prediction patterns and allow the robotic control system to regulate the robot's working pace in accord with those results. The findings can open new avenues to more systematically and appropriately harmonizing the interactions between humans and robots, specifically during joint interactions at construction sites. Future investigations can more deeply examine the resulting patterns of the predicted physiological states in different durations at the individual and group levels and consequently develop physiologically aware mechanisms for enhancing human-robot partnership. Future studies can also work on other perceptive robotic systems dependent on human input for safe and effective operations such as teleoperations and telerobotic. The ability to correctly understand workers' mental and physical conditions and accordingly regulate the controlling parameters of the robot holds a promise in promoting safety and productivity during the control of robots.

References

1. T. Bock, T. Linner, Single-task construction robots by category, in *Constr. Robot*, (Cambridge University Press, Cambridge, 2016), pp. 14–290. https://doi.org/10.1017/CBO9781139872041.002
2. S. Petters, R. Belden, S.A.M., the robotic bricklayer, SMART/Dynamics Mason. 1 (2014) 10–14.
3. R. Sweet, The contractor who invented a construction robot. Constr. Res. Innov. **9**, 9–12 (2018). https://doi.org/10.1080/20450249.2018.1442702
4. Mule 135 Owner's Manual, Mule 135 Owner's Manual, in: 2018. https://www.construction-robotics.com/wp-content/uploads/2019/02/MULE-135-MANUAL.pdf
5. D. Kulić, E. Croft, Pre-collision safety strategies for human-robot interaction. Auton. Robot. **22**, 149–164 (2007). https://doi.org/10.1007/s10514-006-9009-4
6. O. David, F.-X. Russotto, M. Da Silva Simoes, Y. Measson, Collision avoidance, virtual guides and advanced supervisory control teleoperation techniques for high-tech construction: framework design. Autom. Constr. **44**, 63–72 (2014). https://doi.org/10.1016/j.autcon.2014.03.020
7. F. Paas, J.E. Tuovinen, H. Tabbers, P.W.M. Van Gerven, Cognitive load measurement as a means to advance cognitive load theory. Educ. Psychol. **38**, 63–71 (2003). https://doi.org/10.1207/S15326985EP3801
8. K. Schaaff, T. Schultz, Towards an EEG-based emotion recognizer for humanoid robots, Proc. – IEEE Int. Work. Robot Hum. Interact. Commun. 792–796. https://doi.org/10.1109/ROMAN.2009.5326306. (2009)
9. B. Nan, M. Okamoto, T. Tsuji, A hybrid motion classification approach for EMG-based human–robot interfaces using bayesian and neural networks. IEEE Trans. Robot. **25**, 502–511 (2009). https://doi.org/10.1109/TRO.2009.2019782
10. V. Villani, F. Pini, F. Leali, C. Secchi, Survey on human–robot collaboration in industrial settings: Safety, intuitive interfaces and applications. Mechatronics **55**, 248–266 (2018). https://doi.org/10.1016/j.mechatronics.2018.02.009
11. M. Vasic, A. Billard, Safety issues in human-robot interactions. 2013 IEEE Int. Conf. Robot. Autom, 197–204 (2013). https://doi.org/10.1109/ICRA.2013.6630576
12. E. Berger, D. Vogt, S. Grehl, B. Jung, H. Ben Amor, Estimating perturbations from experience using neural networks and Information Transfer, 2016 IEEE/RSJ Int. Conf. Intell. Robot. Syst. **176–181** (2016 November). https://doi.org/10.1109/IROS.2016.7759052

13. N. Bergstrom, T. Kanda, T. Miyashita, H. Ishiguro, N. Hagita, *Modeling of natural human-robot encounters, in: 2008 IEEE/RSJ Int* (Conf. Intell. Robot. Syst, IEEE, 2008), pp. 2623–2629. https://doi.org/10.1109/IROS.2008.4650896
14. A.F. Salazar-Gomez, J. Delpreto, S. Gil, F.H. Guenther, D. Rus, Correcting robot mistakes in real time using EEG signals. Proc. – IEEE Int. Conf. Robot. Autom, 6570–6577 (2017). https://doi.org/10.1109/ICRA.2017.7989777
15. Y. Liu, M. Habibnezhad, H. Jebelli, Brain-computer interface for hands-free teleoperation of construction robots. Autom. Constr. **123**, 103523 (2021). https://doi.org/10.1016/j.autcon.2020.103523
16. Y. Liu, M. Habibnezhad, H. Jebelli, Brainwave-driven human-robot collaboration in construction. Autom. Constr. **124**, 103556 (2021). https://doi.org/10.1016/j.autcon.2021.103556
17. L. Antão, R. Pinto, J. Reis, G. Gonçalves, F.L. Pereira, Cooperative human-machine interaction in industrial environments, 13th APCA Int. Conf. Control Soft Comput. Control. 2018 – Proc. (2018) 430–435. https://doi.org/10.1109/CONTROLO.2018.8514549.
18. S. Shakerian, M. Habibnezhad, A. Ojha, G. Lee, Y. Liu, H. Jebelli, S. Lee, Assessing occupational risk of heat stress at construction: A worker-centric wearable sensor-based approach, Saf. Sci. 142 (2021) 105395. https://doi.org/10.1016/j.ssci.2021.105395
19. V. Villani, F. Pini, F. Leali, C. Secchi, Survey on human–robot collaboration in industrial settings: Safety, intuitive interfaces and applications. Mechatronics **55**, 248–266 (2018). https://doi.org/10.1016/j.mechatronics.2018.02.009
20. H. Jebelli, S. Hwang, S.H. Lee, EEG-based workers' stress recognition at construction sites. Autom. Constr. **93**, 315–324 (2018). https://doi.org/10.1016/j.autcon.2018.05.027
21. H. Li, D. Wang, J. Chen, X. Luo, J. Li, X. Xing, Pre-service fatigue screening for construction workers through wearable EEG-based signal spectral analysis. Autom. Constr **106** (2019). https://doi.org/10.1016/j.autcon.2019.102851
22. F. Lotte, L. Bougrain, A. Cichocki, M. Clerc, M. Congedo, A. Rakotomamonjy, F. Yger, A review of classification algorithms for EEG-based brain–computer interfaces: a 10 year update. J. Neural Eng. **15** (2018). https://doi.org/10.1088/1741-2552/aab2f2
23. M.B.I. Reaz, M.S. Hussain, F. Mohd-Yasin, Techniques of EMG signal analysis: detection, processing, classification and applications. Biol. Proced. Online **8**, 11–35 (2006). https://doi.org/10.1251/bpo115
24. B.C. Jiang, C.A. Gainer, A cause-and-effect analysis of robot accidents. J. Occup. Accid. **9**, 27–45 (1987). https://doi.org/10.1016/0376-6349(87)90023-X
25. A.D. Dragan, S. Bauman, J. Forlizzi, S.S. Srinivasa, Effects of robot motion on human-robot collaboration. Proc. Tenth Annu. ACM/IEEE Int. Conf. Human-Robot Interact. - HRI **15**, 51–58 (2015 March). https://doi.org/10.1145/2696454.2696473
26. M.Y. Leung, Q. Liang, P. Olomolaiye, Impact of job stressors and stress on the safety behavior and accidents of construction workers. J. Manag. Eng. **32**, 1–10 (2016). https://doi.org/10.1061/(ASCE)ME.1943-5479.0000373
27. P.E.D. Love, D.J. Edwards, Z. Irani, Work stress, support, and mental health in construction. J. Constr. Eng. Manag. **136**, 650–658 (2010). https://doi.org/10.1061/(ASCE)CO.1943-7862.0000165
28. P. Antonenko, F. Paas, R. Grabner, T. van Gog, Using electroencephalography to measure cognitive load. Educ. Psychol. Rev. **22**, 425–438 (2010). https://doi.org/10.1007/s10648-010-9130-y
29. F. Paas, J.E. Tuovinen, H. Tabbers, P.W.M. Van Gerven, Cognitive load measurement as a means to advance cognitive load theory. Educ. Psychol. **38**, 63–71 (2003). https://doi.org/10.1027/0227-5910/a000189
30. L. Peternel, N. Tsagarakis, D. Caldwell, A. Ajoudani, Adaptation of robot physical behaviour to human fatigue in human-robot co-manipulation. IEEE-RAS Int. Conf. Humanoid Robot, 489–494 (2016). https://doi.org/10.1109/HUMANOIDS.2016.7803320
31. L. Boccanfuso, Q. Wang, I. Leite, B. Li, C. Torres, L. Chen, N. Salomons, C. Foster, E. Barney, Y.A. Ahn, B. Scassellati, F. Shic, A thermal emotion classifier for improved human-robot

interaction. 25th IEEE Int. Symp. Robot Hum. Interact. Commun. RO-MAN **718–723**, 2016 (2016). https://doi.org/10.1109/ROMAN.2016.7745198
32. A. Rajavenkatanarayanan, H.R. Nambiappan, M. Kyrarini, F. Makedon, Towards a real-time cognitive load assessment system for industrial human-robot cooperation. 29th IEEE Int. Conf. Robot Hum. Interact. Commun. RO-MAN **698–705**, 2020 (2020). https://doi.org/10.1109/RO-MAN47096.2020.9223531
33. Y. Liu, M. Habibnezhad, S. Shayesteh, H. Jebelli, S. Lee, Paving the Way for Future EEG Studies in Construction: Dependent Component Analysis for Automatic Ocular Artifact Removal from Brainwave Signals, J. Constr. Eng. Manag. 147 (2021) 04021087. https://doi.org/10.1061/(ASCE)CO.1943-7862.0002097
34. Y. Liu, M. Habibnezhad, H. Jebelli, S. Asadi, S. Lee, Ocular Artifacts Reduction in EEG Signals Acquired at Construction Sites by Applying a Dependent Component Analysis (DCA), in: Constr. Res. Congr. 2020, American Society of Civil Engineers, Reston, VA, 2020: pp. 1281–1289. https://doi.org/10.1061/9780784482865.135
35. H. Jebelli, S. Hwang, S. Lee, EEG Signal-Processing Framework to Obtain High-Quality Brain Waves from an Off-the-Shelf Wearable EEG Device, J. Comput. Civ. Eng. 32 (2018) 1–12. https://doi.org/10.1061/(ASCE)CP.1943-5487.0000719
36. A. Freedy, E. DeVisser, G. Weltman, N. Coeyman, *Measurement of trust in human-robot collaboration, in: 2007 Int* (Symp. Collab. Technol. Syst, IEEE, 2007), pp. 106–114. https://doi.org/10.1109/CTS.2007.4621745
37. H. Liu, W. Zhou, X. Lai, S. Zhu, An efficient inverse kinematic algorithm for a PUMA560-structured robot manipulator. Int. J. Adv. Robot. Syst. **10** (2013). https://doi.org/10.5772/56403
38. C. Morato, K.N. Kaipa, B. Zhao, S.K. Gupta, Toward safe human robot collaboration by using multiple kinects based real-time human tracking. J. Comput. Inf. Sci. Eng. **14** (2014). https://doi.org/10.1115/1.4025810
39. S. Ma, C. Ren, C. Ye, An omnidirectional mobile robot: Concept and analysis, in *2012 IEEE Int. Conf. Robot. Biomimetics, IEEE*, (2012), pp. 920–925. https://doi.org/10.1109/ROBIO.2012.6491086
40. T. Linner, A. Shrikathiresan, M. Vetrenko, B. Ellmann, T. Bock, Modeling and operating robotic environments using Gazebo/R.O.S. Proc. 28th Int. Symp. Autom. Robot. Constr. ISARC 2011, 957–962 (2011). https://doi.org/10.22260/isarc2011/0177
41. A. Mohammed, L. Wang, Brainwaves driven human-robot collaborative assembly. CIRP Ann. **67**, 13–16 (2018). https://doi.org/10.1016/j.cirp.2018.04.048
42. M. Tanaka, Y. Shigihara, A. Ishii, M. Funakura, E. Kanai, Y. Watanabe, Effect of mental fatigue on the central nervous system: an electroencephalography study. Behav. Brain Funct. **8**, 48 (2012). https://doi.org/10.1186/1744-9081-8-48
43. E. Iáñez, J.M. Azorín, A. Úbeda, J.M. Ferrández, E. Fernández, Mental tasks-based brain-robot interface. Robot. Auton. Syst. **58**, 1238–1245 (2010). https://doi.org/10.1016/j.robot.2010.08.007
44. E. Mower, D.J. Feil-Seifer, M.J. Matarić, S. Narayanan, Investigating implicit cues for user state estimation in human-robot interaction using physiological measurements. Proc. - IEEE Int. Work. Robot Hum. Interact. Commun, 1125–1130 (2007). https://doi.org/10.1109/ROMAN.2007.4415249
45. X. Qu, Effects of cognitive and physical loads on local dynamic stability during gait. Appl. Ergon. **44**, 455–458 (2013). https://doi.org/10.1016/j.apergo.2012.10.018
46. F.G.W.C. Paas, J.J.G. Van Merriënboer, Instructional control of cognitive load in the training of complex cognitive tasks. Educ. Psychol. Rev. **6**, 351–371 (1994). https://doi.org/10.1007/BF02213420
47. Z. Sedighi Maman, M.A. Alamdar Yazdi, L.A. Cavuoto, F.M. Megahed, A data-driven approach to modeling physical fatigue in the workplace using wearable sensors. Appl. Ergon. **65**, 515–529 (2017). https://doi.org/10.1016/j.apergo.2017.02.001

48. A. Pereira, D. Folgado, F. Nunes, J. Almeida, I. Sousa, Using inertial sensors to evaluate exercise correctness in electromyography-based home rehabilitation systems. Med. Meas. Appl. MeMeA 2019- Symp. Proc (2019). https://doi.org/10.1109/MeMeA.2019.8802152
49. A.P. Allen, P.J. Kennedy, S. Dockray, J.F. Cryan, T.G. Dinan, G. Clarke, The Trier Social Stress Test: Principles and practice. Neurobiol. Stress. **6**, 113–126 (2017). https://doi.org/10.1016/j.ynstr.2016.11.001
50. S.R. Allred, L.E. Crawford, S. Duffy, J. Smith, Working memory and spatial judgments: Cognitive load increases the central tendency bias. Psychon. Bull. Rev. **23**, 1825–1831 (2016). https://doi.org/10.3758/s13423-016-1039-0
51. C.C. Morey, N. Cowan, When visual and verbal memories compete: Evidence of cross-domain limits in working memory. Psychon. Bull. Rev. **11**, 296–301 (2004). https://doi.org/10.3758/BF03196573
52. Morgan Quigley, Brian Gerkey, Ken Conley, Josh Faust, Tully Foote, Jeremy Leibs, Eric Berger, Rob Wheeler, and Andrew Ng, ROS: an open-source Robot Operating System, 2009. http://www.willowgarage.com/sites/default/files/icraoss09-ROS.pdf (accessed June 08, 2020).
53. H. Jebelli, M. Mahdi Khalili, S. Lee, A continuously updated, computationally efficient stress recognition framework using electroencephalogram (EEG) by applying Online Multitask Learning Algorithms (OMTL). IEEE J. Biomed. Heal. Informatics. **23**, 1928–1939 (2019). https://doi.org/10.1109/JBHI.2018.2870963
54. S.R. Nandakumar, M. Le Gallo, I. Boybat, B. Rajendran, A. Sebastian, E. Eleftheriou, Mixed-precision architecture based on computational memory for training deep neural networks. 2018 IEEE Int. Symp. Circuits Syst., IEEE, 1–5 (2018). https://doi.org/10.1109/ISCAS.2018.8351656

Chapter 10
Workers' Trust in Collaborative Construction Robots: EEG-Based Trust Recognition in an Immersive Environment

Shayan Shayesteh, Amit Ojha, and Houtan Jebelli

Abstract Today, advances in robotics and autonomous systems enable construction workers to collaboratively work with robots, assigning physically demanding tasks to them. However, working alongside an industrial robot is a novel experience that may take a heavy toll on workers' bodies and minds, particularly in dynamic and uncertain environments, such as construction sites. Given that trust is identified as a critical element for successful cooperation between humans and robots, effective measurement of workers' trust in collaborative robots can lead to practical evaluation of human-robot partnership. In this context, most studies of trust have relied on self-reports. Nevertheless, questionnaires are unable to determine the trust promptly, impeding early preventive interventions. Furthermore, they are invasive, interfering with workers' daily operations. To address these issues, this study proposes a procedure to non-invasively and continuously recognize workers' trust in collaborative construction robots using electroencephalogram (EEG) signals. To that end, an experiment was conducted in which human workers performed a collaborative construction task (i.e., handling materials) with a virtual robot in an immersive environment. Meanwhile, workers' EEG signals were recorded using a wearable sensor. Subsequently, the level of trust of the workers in collaborative robots was measured using the Trust Perception Scale-HRI. By analyzing acquired signals and applying different machine learning algorithms, it was found that EEG signals can be implemented to differentiate levels of trust of construction workers in their robotic counterparts. These findings suggest the feasibility of using workers' EEG signals as a reliable, real-time indicator of trust in collaborative construction robots, which can be regarded as a practical approach for evaluating human-robot collaboration.

S. Shayesteh · A. Ojha · H. Jebelli (✉)
Architectural Engineering Department, Pennsylvania State University, University Park, PA, USA
e-mail: sks6495@psu.edu; azo5242@psu.edu; hjebelli@psu.edu

H. Jebelli et al. (eds.), *Automation and Robotics in the Architecture, Engineering, and Construction Industry*, https://doi.org/10.1007/978-3-030-77163-8_10

Keywords Human-robot collaboration · Electroencephalogram · Trust recognition · Virtual reality · Supervised learning

10.1 Introduction

Robots are becoming predominant in many construction sites in response to established challenges of this industry, including poor productivity, skilled labor shortage, and an aging workforce. Construction robots can perform various operations that are repetitive, exhausting, or undesirable for human workers [1]. Besides, they can securely operate in various circumstances that are often considered hazardous for human beings [2]. However, robot operation in dynamic and unstructured construction jobsites has some downsides. Currently available robots are mostly single-task robots with limited degrees of autonomy. This limitation hinders their vast adoption for many operations in ever-changing construction work environments. Therefore, in many situations, a collaboration in performing construction tasks between humans and robots is required.

This collaboration allows robots to take over physically demanding tasks while enabling humans to mainly focus on operations that require flexibility and dexterity. To ensure a sustainable relationship in human-robot teams, it is vital to understand human factors during interactions and integrate them into developing collaboration strategies. In this context, human trust in robots plays an essential role in the success and efficiency of human-robot teams [3]. Without an adequate amount of trust, robots can be susceptible to either disuse or misuse [4]. In this regard, a lack of trust in collaborative robots can lead to disuse, meaning people are not willing to use them and disregard their capabilities. On the other hand, over-trusting in collaborative robots can lead to excessive reliance on them, causing possible failures in critical situations [5]. Therefore, having commensurate trust in collaborative robots is a must for a successful human-robot interaction at construction sites. Accordingly, reliable measurement of human trust is crucial for evaluating different aspects of human-robot collaboration tasks.

To date, several types of trust measurement techniques have been employed to assess trust in different automated systems, including subjective measures [6], behavioral measures [7], and psychophysiological measures [8, 9]. Prior research has employed self-reporting methods for assessing trust in robots [10]. While these methods provided valuable insights into understanding human trust in automated systems, they lacked the ability to assess trust objectively. Furthermore, such methods were not suitable for developing real-time trust assessment tools. Another approach to inferring trust is based on observing human behavior, which presumes that trust can form the basis for intentions and, eventually, mediate behaviors [11].

Psychophysiological-based approaches can also provide a reliable and objective process for trust assessment, which has the capability of being used in real time [12].

While previous research has explored the feasibility of physiological-based measurement of trust in automated systems, there is a lack of work regarding robots in dynamic construction fields. Given the unique characteristics of construction jobsites, an in-depth understanding of workers' trust in robotic co-workers is still required. To that end, this study aims to investigate a framework to assess human trust in construction robots using continuous electroencephalogram (EEG) recording. To achieve this goal, an EEG-based trust measurement framework was designed and examined through an experimental procedure. Supervised learning models were then implemented to recognize different levels of trust in robots. The results demonstrated the feasibility of using such an approach to continuously measure trust in collaborative construction robots. These findings can be implemented in developing frameworks for evaluating the efficiency of human-robot collaboration at construction jobsites.

10.2 Trust in Human-Robot Collaboration

10.2.1 Trust in Automated Systems

In general, trust can be defined as a firm belief about another's intention or willingness to act by following their words, expressions, decisions, or actions [9]. As humans are increasingly required to interact with intelligent, automated systems, trust becomes an essential factor for synergistic interrelationships. In this context, adequate trust can mediate the relationships between human and automated systems [3]. Increasing levels of automation have caused humans to be displaced as the primary decision-makers [13]. However, trust in technological systems is typically concerned with humans' expectations of the system and how it performs. Thus, trust can be regarded as a property of the human user concerning the system. As a result, trust miscalibration can lead to either misuse or disuse of the system [11].

Performance of an automated system is a major factor that is presumed to have an impact on human trust. Unpredictability is also another factor affecting trust between humans and automated systems [12]. Although a significant body of the existing work on trust in technologies is centered on automated systems [6, 14], work on trust in robotic systems is a relatively newer endeavor. In this regard, trust in automated systems can provide a foundation for understanding trust in robotic systems. Nevertheless, it should be noted that robots may differ from other types of automation, which affect human trust in particular ways that are not yet fully understood. For example, robotic systems tend to be less reliable than accustomed automated systems, given their complexity and dynamic environments in which they are expected to operate [15].

10.2.2 Trust in Collaborative Robots

Today, intelligent robots interact with humans in a sense that demands a greater level of trust [16]. Given that robots are often designed to replace or work with humans in high-risk circumstances, trust is a critical factor that needs to be taken into consideration [3]. In this regard, the physical embodiment of a robot can be a key point in forming trust [4]. It is mainly because people often judge a robot's trustworthiness based on its physical appearance. For instance, robots that have human-like characteristics tend to be considered more trustworthy [17]. Hancock et al. [18] performed a comprehensive analysis of factors affecting trust in robots. This study classified factors affecting trust in human-robot interaction (HRI) into three categories, i.e., human, robot, and environment. Based on this work, human-related factors include training, expertise, situational awareness, and demographic information. Robot-related factors include behavior, dependability, reliability, level of automation, failure rates, false alarms, transparency, as well as attribute-based factors such as type, location, adaptability, and anthropomorphism. Finally, environmental factors include teamwork, culture, communication, shared mental models, task type, task complexity, and multi-tasking.

Researchers have attempted to predict human trust in collaborative robots using dynamic models that rely on the self-reported behavior of humans [7]. However, it may not be practical to use self-reported behavior as a feedback control variable. This is mainly because, similar to subjective ratings, behavioral measures tend to be discrete, making it challenging to trace instant changes in trust. Therefore, alternative methods need to be applied to sense trust levels [19]. Other studies also applied computational models for integrating human trust into robot decision-making. The results showed the feasibility of such models for trust calibration to improve human-robot team performance [5]. In another study, Freedy et al. [20] described a collaborative performance model that captured critical attributes of the human-robot teams and the environment. They found out trust has a critical role in the decision-making process and thereby the effectiveness of human-robot collaboration. Their study also highlighted the benefits of both subjective and objective measures of trust in human-robot collaboration.

10.2.3 Trust Measurement Techniques

Subjective methods of trust measurement in the context of interaction with robots mostly rely on self-reported ratings of individuals. These measurements can be taken before the interaction, during the interaction, or after the interaction. Researchers have employed subjective measures of trust in different domains. Yagoda et al. [3] developed a human-robot interaction (HRI) trust scale, which involved generating a preliminary list of item dimensions within each of the five HRI attributes, namely, team configuration, team process, context, task, and system.

Salem et al. [21] also conducted an experiment to explore factors that affect human perception and trust regarding an erroneous robot. While these methods provided valuable insights into the concept of trust in automated systems and particularly robots, they have certain limitations, such as being disruptive and having low temporal resolution. Furthermore, they cannot be directly used to inform robot decision-making.

To develop intelligent robots that can respond to changes in human trust, real-time sensing of trust levels is essential [13]. To that end, the use of physiological response measurement is motivated by its ability to capture humans' trust in real time. Many researchers have used physiological signals for objective assessment of human trust toward the technology [8]. Akash et al. [16] proposed two approaches for developing classifier-based trust sensor models using EEG and galvanic skin response (GSR) measurements. Their work demonstrated that physiological measurements could be used to estimate human trust in intelligent systems in real-time. Khawaji et al. [22] found that the average of GSR values is significantly affected by both trust and cognitive load in the text-chat environment. Hu et al. [13] proposed and validated a classifier model for sensing individuals' trust in automated systems using electrodermal activity (EDA) and EEG signals. Boudreau et al. [23] found a difference in peak amplitudes of ERP components in human subjects while participating in a coin toss experiment that simulated trust and distrust. Lu and Sarter [11] measured trust in automated systems using eye tracking technology, suggesting that eye tracking is an effective technique for identifying changes in trust levels in real time. The aforementioned studies highlighted the advantages of physiological-based trust measurement in real-world environments.

10.3 EEG-Based Trust Measurement in HRC

10.3.1 Overview

This study proposes a framework for objective assessment of trust in collaborative construction robots using continuously captured physiological signals (i.e., EEG). To that end, first, captured EEG signals were decontaminated from intrinsic and extrinsic artifacts. Then, windowing and feature selection methods were applied to extract measurable parameters (i.e., features) from the signals. The extracted features were labeled as two classes (i.e., high trust level and low trust level) based on the results of the Trust Perception Scale-HRI questionnaire [10]. Afterward, several supervised learning algorithms were employed to classify EEG signals based on the individuals' trust levels. Figure 10.1 demonstrates an overview of the proposed framework. More details about the steps of the study are provided in the following subsections.

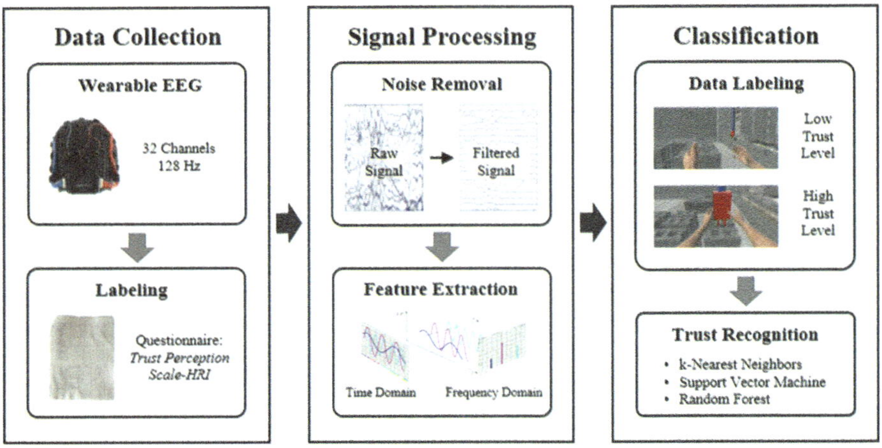

Fig. 10.1 An overview of the proposed framework for EEG-based trust measurement in collaborative construction robots

10.3.2 EEG Signals Artifact Reduction

EEG signals can provide beneficial information regarding human brain activity in different situations [24–27]. These physiological signals can be applied to assess various cognitive states, including trust, during human-robot collaboration tasks. However, EEG signals can be contaminated with artifacts due to several factors (e.g., environmental noises, facial muscle movements) that reduce the quality of the captured signals. These artifacts can be generated by internal sources in the human body (intrinsic artifacts) or produced by external sources (extrinsic artifacts). To obtain high-quality EEG signals, effective artifact removal techniques must be applied to noisy signals [28]. In this study, a fixed-gain filtering method was employed to reduce extrinsic artifacts of the EEG signals. Notably, a bandpass filter with a low cutoff frequency of 0.5 Hz and a high cutoff frequency of 45 Hz was applied to raw EEG signals. To reduce intrinsic artifacts, the independent component analysis (ICA) method was employed to remove artifactual components of EEG signals. Scalp maps pertinent to each channel of the EEG device were generated and inspected for components representing artifacts (e.g., muscular movement, blinking, eye movement). Afterward, artifactual components were removed, and clean EEG signals were acquired.

10.3.3 EEG Signals Feature Extraction

After reducing extrinsic and intrinsic EEG signal artifacts, useful features related to human trust were extracted from the clean signals. Features are informative and measurable properties of the signals that can be obtained in both time and frequency domains [29]. In this study, 12 features in time and frequency domains were selected based on the literature to detect EEG signal patterns (Table 10.1). To calculate each feature, EEG signals were divided into several windows with a size of 1 s (consisted of 128 data points), as feature calculation over single EEG data points is not informative. The fixed windowing approach was selected in this study, and the window size was determined based on preceding studies in the literature. Features were then calculated from these windowed segments for each EEG device channel to be implemented in machine learning models.

10.3.4 Supervised Learning Algorithms for Trust Assessment

In this study, different supervised machine learning algorithms were applied to assess different trust levels in collaborative construction robots. The selected algorithms included k-nearest neighbors (k-NN), support vector machine (SVM), and random forest. k-NN was the first algorithm that was applied to the extracted features. It is a memory-based algorithm that uses the entire database for prediction, according to a similarity measure in the instance space [30]. This algorithm can make predictions by creating a group of nearby data points with similar features, known as neighbors. This algorithm operates based on the assumption that

Table 10.1 Selected features in time and frequency domains and their explanation

	Feature	Explanation
Time domain	Mean value	Average absolute value of amplitude
	Standard deviation	Variation of signals
	Peak	Maximum value of amplitude
	Peak to peak	Time between peaks between the various windows
	Kurtosis	Shows the sharpness of signal peak
	Zero cross number	Number of points where the sign of amplitude changes
	Median value	Median of the signal
Frequency domain	Delta band power	Power of the signal in delta frequency range
	Theta band power	Power of the signal in theta frequency range
	Alpha band power	Power of the signal in alpha frequency range
	Beta band power	Power of the signal in beta frequency range
	Median frequency	Half of the signal power is distributed in less frequencies

neighbors in the instance space have the same class [31]. k-NN measures the distance between the unlabeled target data points and labeled training data points using Eq. 10.1:

$$d_E\left(X_i,Z_j\right)=X_i-Z_j=\sqrt{\left(\sum_{l=1}^{d}\left(x_i^l-z_j^l\right)^2\right)} \tag{10.1}$$

where $d_E(X_i, Z_j)$ is the distance between unlabeled data points (X_i) and labeled data points (Z_i). Afterward, a subset of k nearest neighbors to unlabeled data point ($\theta_k(X_i) = \{\theta_1, \theta_2, \theta_3, \ldots, \theta_k\} \subseteq \{Z_1, Z_2, Z_3, \ldots, Z_n\}$) and the class labels ($L(\theta_k) = y$ where $y \in \{+1$(high trust level), -1(low trust level)$\}$) would be derived. Subsequently, the predicted label for X_i can be defined as the majority labels of $\theta_k(X_i)$ using Eq. 10.2:

$$L\left(X_i\right)=\begin{cases}+1 \text{ high trust level if } \left|A_{+1}\right|\geq\left|A_{-1}\right| \\ -1 \text{ low trust level if } \left|A_{+1}\right|<\left|A_{-1}\right|\end{cases} \tag{10.2}$$

where A_{+1} and A_{-1} are sets of neighbor data points labeled as +1 (high trust level) and −1 (low trust level), respectively.

The authors also applied the SVM algorithm, which can be regarded as an appropriate classifier for physiological data [32]. This algorithm generates hyperplanes that classify data points of a binary classification problem. By minimizing the following objective function, SVM delivers an optimal hyperplane that maximizes the margin between data points of two classes:

$$\min_{w,b,\xi}\frac{1}{2}w^2+\frac{c}{n}\sum_{i}^{n}\xi_i,\quad for \quad i=1,\ldots,N \tag{10.3}$$

where c is a user-specified tuning parameter and n is the number of training data points. Linear SVM can solve the objective function given the following constraints [$y_i(<w, x> + b) \geq 1 - \xi_i, i = 1, \ldots, n, \xi_i \geq 0$]. Non-linear SVM can also solve the objective function given the following constraints:

$$y_i\left(\sum_{j=1}^{m}w_i\phi_j\left(X_i\right)+w_0\right)\geq 1-\xi_i,\quad i=1,\ldots,n,\quad \xi_i\geq 0 \tag{10.4}$$

where X_i is a data point, y_i is the class label, w is the weight vector, $\phi(X_i)$ is the mapping function, and ξ_i is a nonnegative slack variable that generalizes a classifier with soft margins [26]. By selecting the proper mapping function, the class of unlabeled data can be defined using Eq. 10.5:

$$L\left(X_i\right)=sign\left(w^T\phi\left(X_i\right)\right)=\begin{cases}+1 \text{ high trust level} \\ -1 \text{ low trust level}\end{cases} \tag{10.5}$$

The last algorithm that was tested is the random forest. It employs an ensemble learning method for classification using multiple decision trees during the training process and yields average prediction of individual trees [33]. This algorithm creates forests with random amounts of decision trees that are created from the training samples. In this case, various variations of the main classification can be achieved. Random decision trees are generated, namely, $Tree_1$, $Tree_2$, …, $Tree_n$, whose corresponding outputs are k_1, k_2, …, k_n. Majority voting is undertaken in this method, and the output of the classifier algorithm is the class k with the majority of the votes [33].

10.4 Experimental Procedure

To examine the performance of the proposed framework for assessing human trust in collaborative construction robots, an experiment was designed and conducted in an immersive environment. To that end, a virtual construction site was simulated using the Unity game engine to enable safe and controlled interactions with virtual robots. Based on authors' previous study [34], an integrated system of tracking devices was employed to mimic realistic movements and interaction mechanisms of actual human-robot collaboration in a simulated environment. The virtual robot movements were also modeled through an inverse kinematics approach using in-house scripts for the game engine. Two scenarios of a masonry task were designed to induce different levels of trust in collaborative robots. The first scenario was to work along with an autonomous version of the material lifting robot. Since humans could not be aware of the robot's movements in this scenario, it was assumed that this scenario would result in a lower level of trust in the robot. The second scenario was to work along with a semi-autonomous version of the robot, of which humans had control over most of its movements.

Seven healthy subjects participated in the experimental process. Subjects were between 25 and 35 years old (M = 27.7, SD = 3.5), and all were males. None of the subjects had any medical condition that could affect their task performance in this study. A consent form and a description of the procedure were provided to subjects prior to the experiment. Afterward, they were equipped with a wearable EEG device (Emotiv EPOC Flex) and a VR headset (HTC VIVE Pro Eye). The EEG device featured 32 channels positioned on the subjects' scalps and was connected wirelessly to a nearby computer to record brain waves with a rate of 128 Hz. Once the experimental setup was prepared, subjects were asked to perform the masonry task in collaboration with a virtual construction robot. Subjects performed both scenarios, with 15 min rest in between, in random orders to mitigate the learning effect. Subjects' brain activities were being captured continuously via EEG device throughout 10-min task performances.

After completing each scenario, subjects were asked to indicate their trust level in the construction robot by filling out a Trust Perception Scale for Human-Robot Interaction (HRI) questionnaire (Table 10.2). A 14-item version of the self-report scale was employed in this study, scored based on a percentage (from 0 to 100). The

Table 10.2 Trust Perception Scale-HRI and its items, based on [10]

	Trust Perception Scale-HRI
	What % of the time will this robot be...
1	Dependable
2	Reliable
3	Unresponsive
4	Predictable
5	Friendly
	What % of the time will this robot...
6	Act consistently
7	Malfunction
8	Have errors
9	Provide feedback
10	Meet the needs of the task
11	Provide appropriate information
12	Communicate with people
13	Perform exactly as instructed
14	Follow directions

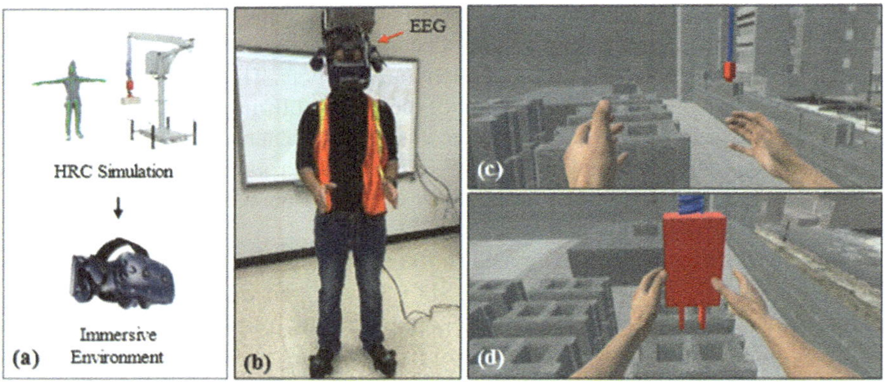

Fig. 10.2 Experimental procedure: (**a**) simulating human-robot collaboration in an immersive setting, (**b**) subject performing the task, (**c**) low trust level task, (**d**) high trust level task

average of the scores for each item was then calculated to acquire an overall trust score for each subject and each scenario. It should be noted that the scores on items 3, 7, and 8 of the scale were reversed. Subjects' overall scores for both scenarios were then compared and used as labels (i.e., high trust level and low trust level) for training supervised learning models. Figure 10.2 demonstrates the experimental procedure of this study.

10.5 Results and Discussion

Following the data collection process, the data were labeled based on the Trust Perception Scale questionnaire results. The average of the scores for each item was used as the overall trust score. The results indicated that working alongside an autonomous robot would lead to lower levels of trust, while working alongside a semi-autonomous robot with a certain control over it would lead to higher trust levels among humans (Fig. 10.3). The Kruskal-Wallis test was also performed on the average scores to determine any statistically significant differences. The test confirmed a significant difference between trust levels in collaborative robots between different scenarios (p-value = 0.015 < 0.05).

After data labeling, the authors applied different supervised learning algorithms (i.e., k-NN, SVM, and random forest) on the collected data from the experiment. To that end, the collected data were randomly divided into 80% for the training set and 20% for the testing set. The models were trained on the training set, and their accuracy was determined on the testing set. The classification accuracy was calculated as the proportion of the correctly predicted results among the total number of the tested data points. Achieved classification accuracies for implemented algorithms are presented in Table 10.3. As shown in the table, the k-NN algorithm demonstrated the highest prediction accuracy of about 88% among the tested supervised learning algorithms. To further investigate this algorithm, the authors examined different numbers of neighbors (i.e., k). Among different values, k = 3 resulted in the highest prediction accuracy (Fig. 10.4).

The results of this study indicated the capability of the proposed procedure in recognition of construction workers' trust in collaborative robots using EEG signals recorded from a wearable EEG device. This is a promising finding since the

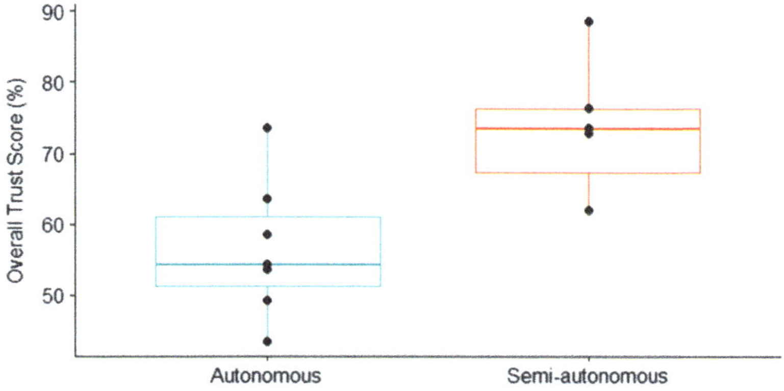

Fig. 10.3 Trust Perception Scale score for two different scenarios (n = 7)

Table 10.3 Achieved prediction accuracies for supervised learning methods

Supervised learning algorithm	Accuracy (%)
k-nearest neighbors (k-NN)	87.9
Support vector machine (SVM)	76
Random forest	75

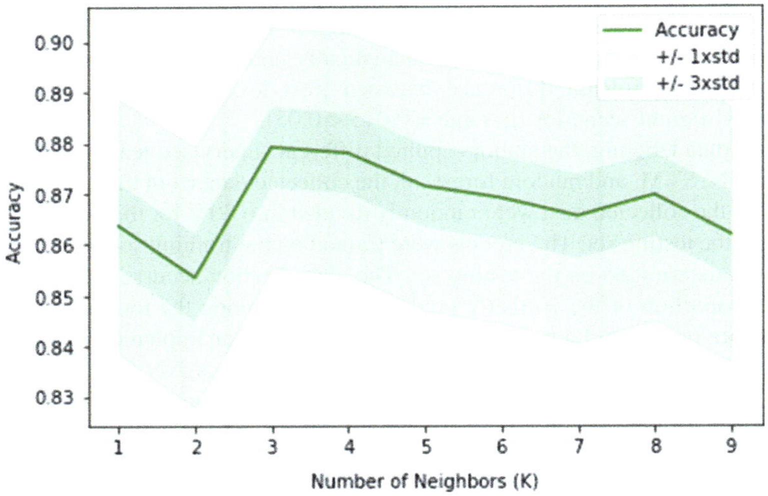

Fig. 10.4 Accuracy of the k-NN algorithm for different numbers of neighbors

previous studies that employed EEG signals for trust assessment in other domains did not take into account the demanding construction tasks and the dynamic nature of construction jobsites. Given that having adequate levels of trust in robots is one of the main aspects of a successful collaboration between humans and robots, the proposed approach can significantly help evaluate the quality and efficiency of such collaboration.

10.6 Conclusion

This study proposed a framework for the non-invasive and continuous measurement of trust in collaborative construction robots using physiological signals (i.e., EEG) and supervised learning algorithms. Through an experimental procedure in an immersive virtual environment, the performance of the proposed framework was examined. The results highlighted the feasibility of using EEG signals as a reliable indicator of trust in dynamic and unstructured construction workplaces. This research contributes to the current body of knowledge by proposing an objective approach for continuous measurement of trust in robotic co-workers, which can be effectively used to evaluate the efficiency in human-robot teams. This study can also pave the path for evaluating other aspects of human-robot collaboration at construction jobsites. In this regard, future research must go beyond and explore other aspects of such collaborations.

References

1. C.-J. Liang, V.R. Kamat, C.C. Menassa, Teaching robots to perform quasi-repetitive construction tasks through human demonstration. Autom. Constr. **120**, 103370 (2020). https://doi.org/10.1016/j.autcon.2020.103370
2. S. You, J.-H. Kim, S. Lee, V. Kamat, L.P. Robert, Enhancing perceived safety in human–robot collaborative construction using immersive virtual environments. Autom. Constr. **96**, 161–170 (2018). https://doi.org/10.1016/j.autcon.2018.09.008
3. R.E. Yagoda, D.J. Gillan, You want me to trust a ROBOT? The development of a human–robot interaction trust scale. Int. J. Soc. Robot. **4**, 235–248 (2012). https://doi.org/10.1007/s12369-012-0144-0
4. B.C. Kok, H. Soh, Trust in Robots: Challenges and opportunities. Curr. Robot. Rep. **1**, 297–309 (2020). https://doi.org/10.1007/s43154-020-00029-y
5. M. Chen, S. Nikolaidis, H. Soh, D. Hsu, S. Srinivasa, Trust-aware decision making for human-robot collaboration. ACM Trans. Hum. Robot Inter. **9**, 1–23 (2020). https://doi.org/10.1145/3359616
6. J.-Y. Jian, A.M. Bisantz, C.G. Drury, Foundations for an empirically determined scale of trust in automated systems. Int. J. Cogn. Ergon. **4**, 53–71 (2000). https://doi.org/10.1207/S15327566IJCE0401_04
7. D. Miller, M. Johns, B. Mok, N. Gowda, D. Sirkin, K. Lee, W. Ju, Behavioral measurement of trust in automation. Proc. Hum. Fact. Ergonom. Soc. Ann. Meet. **60**, 1849–1853 (2016). https://doi.org/10.1177/1541931213601422
8. I.B. Ajenaghughrure, S.C. Sousa, I.J. Kosunen, D. Lamas, Predictive Model to Assess User Trust, in: Proceedings of the 10th Indian Conference on Human-Computer Interaction, ACM, New York, 2019: pp. 1–10. https://doi.org/10.1145/3364183.3364195
9. K. Gupta, R. Hajika, Y.S. Pai, A. Duenser, M. Lochner, M. Billinghurst, Measuring human trust in a virtual assistant using physiological sensing in virtual reality. In: 2020 IEEE Conference on Virtual Reality and 3D User Interfaces (VR), IEEE, 2020: pp. 756–765. https://doi.org/10.1109/VR46266.2020.00099
10. K.E. Schaefer, Measuring trust in human robot interactions: Development of the "Trust Perception Scale-HRI," in: Robust Intelligence and Trust in Autonomous Systems, Springer US, Boston, 2016: pp. 191–218. https://doi.org/10.1007/978-1-4899-7668-0_10
11. Y. Lu, N. Sarter, Eye tracking: A process-oriented method for inferring Trust in Automation as a function of priming and system reliability. IEEE Trans Hum Mach Syst **49**, 560–568 (2019). https://doi.org/10.1109/THMS.2019.2930980
12. S. Shahrdar, L. Menezes, M. Nojoumian, A survey on trust in autonomous systems. In: 2019: pp. 368–386. https://doi.org/10.1007/978-3-030-01177-2_27
13. W.-L. Hu, K. Akash, N. Jain, T. Reid, Real-time sensing of trust in human-machine interactions. IFAC-PapersOnLine **49**, 48–53 (2016)
14. F. Ekman, M. Johansson, J. Sochor, Creating appropriate Trust in Automated Vehicle Systems: A framework for HMI design. IEEE Trans. Hum.-Mach. Syst. **48**, 95–101 (2018). https://doi.org/10.1109/THMS.2017.2776209

15. A.L. Baker, E.K. Phillips, D. Ullman, J.R. Keebler, Toward an understanding of trust repair in human-robot interaction. ACM Trans Interact. Intell. Syst. **8**, 1–30 (2018). https://doi.org/10.1145/3181671
16. K. Akash, W.-L. Hu, N. Jain, T. Reid, A classification model for sensing human trust in machines using EEG and GSR. ACM Trans. Interact. Intell. Syst. **8**, 1–20 (2018). https://doi.org/10.1145/3132743
17. M. Natarajan, M. Gombolay, Effects of Anthropomorphism and Accountability on Trust in Human Robot Interaction, in: Proceedings of the 2020 ACM/IEEE International Conference on Human-Robot Interaction, ACM, New York, 2020: pp. 33–42. https://doi.org/10.1145/3319502.3374839
18. P.A. Hancock, D.R. Billings, K.E. Schaefer, J.Y.C. Chen, E.J. de Visser, R. Parasuraman, A meta-analysis of factors affecting Trust in Human-Robot Interaction, human factors. J. Hum. Fact. Ergonom. Soc. **53**, 517–527 (2011). https://doi.org/10.1177/0018720811417254
19. R. Riedl, A. Javor, The biology of trust: Integrating evidence from genetics, endocrinology, and functional brain imaging. J. Neurosci. Psychol. Econ. **5**, 63–91 (2012). https://doi.org/10.1037/a0026318
20. A. Freedy, E. DeVisser, G. Weltman, N. Coeyman, Measurement of trust in human-robot collaboration. In: 2007 International Symposium on Collaborative Technologies and Systems, IEEE, 2007: pp. 106–114. https://doi.org/10.1109/CTS.2007.4621745
21. M. Salem, G. Lakatos, F. Amirabdollahian, K. Dautenhahn, Would You Trust a (Faulty) Robot?, in: Proceedings of the Tenth Annual ACM/IEEE International Conference on Human-Robot Interaction, ACM, New York, 2015: pp. 141–148. https://doi.org/10.1145/2696454.2696497
22. A. Khawaji, J. Zhou, F. Chen, N. Marcus, Using Galvanic Skin Response (GSR) to Measure Trust and Cognitive Load in the Text-Chat Environment, in: Proceedings of the 33rd Annual ACM Conference Extended Abstracts on Human Factors in Computing Systems - CHI EA '15, ACM Press, New York, 2015: pp. 1989–1994. https://doi.org/10.1145/2702613.2732766
23. C. Boudreau, M.D. McCubbins, S. Coulson, Knowing when to trust others: An ERP study of decision making after receiving information from unknown people. Soc. Cogn. Affect. Neurosci. **4**, 23–34 (2009). https://doi.org/10.1093/scan/nsn034
24. M. Soleymani, S. Asghari-Esfeden, M. Pantic, Y. Fu, Continuous emotion detection using EEG signals and facial expressions, Proceedings - IEEE International Conference on Multimedia and Expo. 2014-Septe (2014) 1–6. https://doi.org/10.1109/ICME.2014.6890301
25. S. Hwang, A.M. Asce, H. Jebelli, S.M. Asce, B. Choi, M. Choi, S. Lee, M. Asce, Measuring workers' emotional state during construction tasks using wearable EEG. J. Constr. Eng. Manag. **144** (2018). https://doi.org/10.1061/(ASCE)CO.1943-7862.0001506
26. H. Jebelli, S. Hwang, S.H. Lee, EEG-based workers' stress recognition at construction sites. Autom. Constr. **93**, 315–324 (2018). https://doi.org/10.1016/j.autcon.2018.05.027
27. H. Jebelli, M.M. Khalili, S. Lee, Mobile EEG-based workers' stress recognition by applying deep neural network, in *Advances in Informatics and Computing in Civil and Construction Engineering*, (Springer, 2019), pp. 173–180. https://doi.org/10.1007/978-3-030-00220-6_21
28. Y. Liu, M. Habibnezhad, H. Jebelli, Brain-computer interface for hands-free teleoperation of construction robots. Autom. Constr. **123**, 103523 (2021). https://doi.org/10.1016/j.autcon.2020.103523
29. Y. Liu, M. Habibnezhad, H. Jebelli, Brainwave-driven human-robot collaboration in construction. Autom. Constr. **124**, 103556 (2021). https://doi.org/10.1016/j.autcon.2021.103556
30. N.S. Altman, An introduction to kernel and nearest-neighbor nonparametric regression. Am. Stat. **46**, 175 (1992). https://doi.org/10.2307/2685209
31. K. Beyer, J. Goldstein, R. Ramakrishnan, U. Shaft, When is "nearest neighbor" meaningful? in *International Conference on Database Theory*, (1999), pp. 217–235. https://doi.org/10.1007/3-540-49257-7_15
32. L. Changchun, P. Rani, N. Sarkar, An empirical study of machine learning techniques for affect recognition in human-robot interaction, in *2005 IEEE/RSJ International Conference*

on Intelligent Robots and Systems, (IEEE, 2005), pp. 2662–2667. https://doi.org/10.1109/IROS.2005.1545344
33. D.R. Edla, K. Mangalorekar, G. Dhavalikar, S. Dodia, Classification of EEG data for human mental state analysis using random Forest classifier. Procedia Comp. Sci. **132**, 1523–1532 (2018). https://doi.org/10.1016/j.procs.2018.05.116
34. M. Habibnezhad, S. Shayesteh, H. Jebelli, J. Puckett, T. Stentz, Comparison of ironworker' fall risk assessment systems using an immersive biofeedback simulator. Autom. Constr. **103471** (2020). https://doi.org/10.1016/j.autcon.2020.103471

Index

A
Above ground level (AGL), 134
Adapting automation to man, culture and society (quotes), 106
Additive manufacturing (AM), 11
Advanced construction technologies, 5
AECO industry, 4, 8
AECO-related programs, 33
AI integration, 108
AlphaBeta, 33
Applications of automation and robotics technologies
  automating construction process, 17
  demolition/material sorting, 21, 22
  earthwork operations, 18, 19
  health and safety monitoring, 20
  inspection and maintenance, 19, 20
  inventory and stock management, 21
  jobsite surveying, 19
  lifecycle of construction projects, 16
  off-site construction, 18
  on-site fabrication and operations, 17, 18
  real-time decision-making, 20, 21
Architecture, engineering, and construction (AEC), 60, 144
Architecture, engineering, construction, and operations (AECO), 2
  construction technology, 32
  curriculum, 33
  digital transformation, 34
  education, 33
Artificial intelligence (AI), 13, 108, 149
Artificial neural network (ANN), 184
Augmented reality (AR), 10, 114
Autodesk and the Associated General Contractors (AGC), 32
Automated and robotic on-site factories, 18
Automated built environment, 109
Automated construction, 22
Automated robotic systems, 108
Automated sorting system, 101, 103
Automated system, 109
Automated technology, 106
Automatic systems, 74
Automation, 6, 7, 166
  evolution, 76
  integral role, 75
  large-scale implementation, 75
  robotics and reproduction, 74
  sustainability goals, 75
Automation and robotics KSAs
  assessment framework (*see* Competency benchmark and performance assessment framework)
  competency groups, 39–43
  content analysis, 38
  foundational, 39
  literature review, 38
  review, 39
Automation and robotics technologies
  advancements, 2, 15
  applications (*see* Applications of automation and robotics technologies)
  construction industry, 7
  construction workflows, 15
  continuous evaluation, 2
  definition, 6

H. Jebelli et al. (eds.), *Automation and Robotics in the Architecture, Engineering, and Construction Industry*, https://doi.org/10.1007/978-3-030-77163-8

Automation and robotics technologies (*cont.*)
deployment, 2, 16
groups, 22
implementation, 5, 15
Industry 4.0, 7
intelligent systems, 12–15
manufacturing industry, 7
potential capacity, 4
reality computing, 8–9
selection criteria, 15
strategic solution, 7
visualization, 9–11
Automation in construction
contour crafting, 105
economic evaluation, 99
image processing (*see* Image processing, SUD)
perspectives, 99
social aspect, 105, 106
Automotive industry, 98
Autonomous equipment, 12, 18, 19

B
Barriers of construction automation
eliminating Jobs, 88, 89
high initial costs, 88
lack of skilled individuals, 93
less systematic guidance, 89
widespread consensus, 87
Beyond visual line of sight (BVLOS) operations, 130
Big Data (BD), 13
Big data analytics, 41
Building component manufacturing (BCM), 18
Building information modeling (BIM), 4, 5, 116
Built robotics, 19

C
CAD software, 4
Camera integration, 102
Capitalism, 107
Cartesian coordination system, 105
Center for Construction Research and Training (CPWR), 170
Centralized management, 107
Civil infrastructure projects, 62
application categories, 62
human-robot interaction, 62
implementation cost, 67
prioritized project areas, 64
productivity improvement, 62, 67
RAC adoption, 62
reliability analysis, 63
survey design and analysis, 62
survey responses, 63
target population, 63, 65
Classical economic models, 107
Closed-circuit television (CCTV), 116
Cloud-based platforms, 13
Cloud computing, 13, 39
Cognitive load theory (CLT), 186
Combinative multicriteria assessment metrics, 47, 55
Combinative multicriteria evaluation metrics
attitudes competency group, 50, 54
computer competency group, 49
computer science competency group, 51
consensus measurement, 38
round 1, 45
round 2, 45
technology competency group, 49, 52
Comfortable approach distance, 123, 128, 136
Communication and tracking (CT) systems, 14
Communications through digitization, 80, 82
Competency benchmark and performance assessment framework
automation and robotics curriculum, 53
benchmarks' ratings, 53, 54
competency sub-groups, 55
consistency and effectiveness, 54
Delphi study, 55
developing ratings, 54
feedback and recommended modifications, 55
performance and productivity levels, 56
research priorities, 56
resource and guideline, 53, 56
structured work experience, 54
training levels, 53
training program, 54
Computer science, 41, 108
Computing competency, 39
Conceptual BIM, 5
Connected BIM, 5
Construction, 7, 22
AECO, 34
future work, 33
industrialization and digitization, 33
job categories, 34
managers and supervisors, 33
Construction accuracy, 34
Construction and demolition waste (C&D), 99
Construction automation, 7, 61
chain of enablers, 91
contributions, 93

human errors, 93
research, 88
transition, 93
Construction companies, 32
Construction complexity, 60
Construction industry
automation and robotics, 2, 60, 98
challenges, 3, 4
cutting-edge equipment and systems deployment, 2
digital technologies, 98
GDP, 60
global expenditure, 2
productivity issues, 61
robotics, 98
technologically passive, 98
Construction Industry Institute (CII), 170
Construction jobsites
activities, 144
applications, 144
attentional cost, 158
construction work, 149, 151
contemporary construction environments, 148–150
drones, 144
hardware and software capabilities, 144
health, 145, 158
human labor-driven industry, 145
industrial sector, 145
industry, 144, 145
infrastructure, 145
physical risk, 158
potential future uses, 148–150
psychological impact, 159
roadmap, 158
safety, 158
gross domestic product (GDP), 146
industrial sectors, 146
industrial workers, 146
occupational fatalities 2018, 147
personal and societal distress, 146
physical demands, 148
risk levels, 146
sector, 148
US Bureau of Labor Statistics, 146
workers, 147
safety of workers, 158
safety risks, 145
sector, 158
unmanned aerial vehicles (UAVs), 144
Construction management education, 32, 33
Construction managers, 33
Construction productivity, 60, 61, 67
Construction professionals, 64
Construction-related businesses, 5
Construction robots, 7
Construction technology and innovation, 55
Construction wastes, 83
Content analysis-based review method, 35
Contour crafting, 105
Controllability, 74, 75, 77, 80, 84, 93
Conventional construction methods, 73
Curriculum integration, 33

D

Data-driven optimization, 20, 21
Degree of industrialization, 74, 75
Delphi study
consensus, 36, 37, 51
construction technology, 55
final evaluation and rating, 47, 48, 50
first evaluation phase, 37
5-point Likert scale, 38
initial evaluation and rating, 43, 44, 47
iterative evaluation procedure, 36
KSAs, 52
semi-structured interviews, 36, 51, 53
subject matter experts, 36
survey development, 38
Digitalization, 2, 3, 5, 15, 81
Digital technology, 5
Digital twins (DT), 10, 11
Drones and autonomous vehicles, 62, 64

E

Early consensus (EC), 37
Early strong agreement (ESA), 37
Earthwork operations, 18, 19
Economic contributions, construction automation
decreased labor costs, 78–80
decreased material costs, 77, 78
improved communications through digitization, 80, 82
increased productivity, 80, 81
productivity, 77
Electrocardiogram (ECG/EKG) trackers, 14
Electroencephalogram (EEG), 127, 203
artifact reduction, 206
collaborative construction robots, 206
feature extraction, 207
physiological signals, 205
supervised learning algorithms, 207–209
Trust Perception Scale-HRI questionnaire, 205
Emerging 5G network, 98

Emerging automated technologies, 109
Emerging technologies, 84
Employment, 108
End-to-end applications, 6
Engineering ethics, 106
Enhancing construction automation
cost-effectiveness, 92
mass production, 90
modular construction, 90, 91
prefabrication, 90, 91
promoting on-site use, 90–92
Environmental contributions, construction automation
shrinking project duration, 84
social and economic benefits, 82
waste minimization, 82, 84
Era of industrialization, 4–7, 18, 22, 33
Era of Optimization, 5
Eras of construction disruption, 4, 5
Error-related potential (ErrP), 184
Essential KSAs
Delphi method (*see* Delphi study)
job-task performance criteria, 36
literature review, 34
proficiency, 35
semi-structured interviews, 35
European Union Aviation Safety Agency (EASA), 134
Exoskeletons, 62, 64, 168, 169

F
Fault-tolerant mechanism
automized machinery-dominated process, 182
biosensors, 183
biosignals, 183
collaboration tasks, 189
collaborative robots, 183
emerging sensing approaches, 183
experimental tasks, 190
feasibility, 184, 189
human physiology, 184
human-populated environments, 182
human-robot communication strategies, 183
interactions, 182
machine learning (ML), 183
multi-faceted design, 182
N-digit task, 189
performance, 194–196
physical and mental abilities, 182
requirements, 182
robotic operating system (ROS), 189
safety issues, 182
semi-automated mason (SAM), 182
sensory data, 183
single-task construction robots (STCRs), 182
structures, 190
time-variant characteristics, 183
transform construction, 182
UGV's robotic systems, 184
unmanned ground vehicle (UGV), 189
wearable EMG sensors, 189
Federal Aviation Administration (FAA), 134, 151
FLx ErgoSkeleton, 169

G
Galvanic skin response (GSR) measurements, 205
Generative design, 13, 14
Global navigation satellite systems (GNSSs), 115
Globalization, 2
Google Glasses, 10
Government monetary policy, 108
Gross domestic product (GDP), 60

H
Heavy equipment, 18
Human-computer (HCI) interactions, 120
Human-drone communication, 121, 124, 132
Human-drone interaction (HDI)
advantages, 112
aerial construction, 117
aerial platforms, 112
aerial systems, 112
applications, 112
barriers, 131
brain control, 127
challenges, 131, 132
civilian operations, 112
collaboration, 112
commercial open-source drone systems, 132
construction field, 133
construction industry, 112, 113
construction jobsites, 113
construction phase
inspection, 116
mapping and earthwork volumetrics, 115
progress monitoring, 115, 116
safety monitoring, 116

control modalities, 112
controlled and safe environments, 133
definition, 120, 121
distance, 128, 129
elements, 113
environment, 135
factors, 112, 113
gaze interaction, 126
gesture, 124
ground control stations, 132
hardware components, 133
implementation challenges, 113
influential effects, 133
interaction modalities, 123, 124
material handling, 117
multidisciplinary research, 132
natural user interfaces, 112, 133
performance and psychological effects, 133
physical characteristics, 133
post-construction phase
marketing, 120
post-disaster assessment, 119, 120
structure and infrastructure inspection, 118, 119
pre-construction
site evaluation, 114
site planning, 114
procedures, 112
regulations, 134, 135
research areas, 131
roles of drones, 122
roles of humans, 121
security surveillance, 116, 117
site communication, 118
speech, 125, 126
technologies, 112
touch, 127, 128
trust, 129, 130
Human-robot collaboration (HRC)
attribute-based factors, 204
automated systems, 203
autonomy, 202
capability, 211
construction sites, 202
decision-making process, 204
dexterity, 202
dynamic and unstructured construction jobsites, 202
dynamic construction fields, 203
EEG-based HRC masonry task, 191, 192
electroencephalogram (EEG), 203
EMG-based HRC masonry task, 193
experimental procedure, 209, 210
fault-tolerant algorithm, 186, 188, 189
flexibility, 202
human workers, 202
industry, 202
investigations, 184, 185
k-NN algorithm, 212
psychophysiological-based approaches, 203
representative structure, 185, 187
self-reported behavior, 204
supervised learning algorithms, 211
supervised learning methods, 212
trust measurement techniques, 202, 204, 205
Trust Perception Scale score, 211
Human-robot interaction (HRI), 98, 105, 120
Hypothalamus-pituitary-adrenal (HPA) system, 157

I
Image processing, SUD
automatic sorting process, 100
C&D recycling, 99–101
C&D waste, 100
control system, 101
data, 102
economic impact, 104
heavy machineries, 100
high-end technology, 100
hydration reaction, 100
LED, 103
managerial perspective, 103
microcontroller, 102
object detection, 102
professional operator, 103
RGB, 103
rough analysis, 102
schematic protocol, 101
smart recycling plant, 100
system accuracy, 102
system adjustment, 102
unit, code, 103
Independent component analysis (ICA) method, 206
Industrial revolution, 109
Industrialization, 2
Industry 4.0 technologies, 7
Information, 13
Innovative technologies, 4
Intelligent automated systems, 106
Intelligent systems
AI, 13
BD, 13
cloud computing, 13
CT systems, 14

Intelligent systems (*cont.*)
  generative design, 13, 14
  IoT, 12
  wearable sensors, 14
Interaction modalities, 121, 123–128, 133, 136
Internet of Things (IoT), 12, 149
Interquartile range (IQR), 36
Inventory and stock management, 21

J
Jobsite surveying, 19

K
k-nearest neighbors (k-NN), 207
Knowledge, skills, and abilities (KSAs), 32
KSA's construction management (CM), 32
KSAs competency groups
  attitudes competency, 43
  computer science competency, 41
  computing competencies, 39
  sub-groups, 41
  technology competency, 42

L
Labor shortage, 32
Large-scale prefabrication (LSP), 18
Laser scanning, 8, 115
Levels of agreements (LOAs), 8, 37, 38, 43, 44, 47
Light detection and ranging (LiDAR) devices, 115
Load, haul, and dump (LHD), 19
Low accessibility, 87

M
Machine learning, 41, 127
Manufacturing industrialization, 74
Massive lost productivity, 74
Material unit lift enhancer (MULE), 11, 182
Microsoft HoloLens, 10
Mixed reality (VR), 10
Modern post-Keynesian economics, 107
Monetary and fiscal policies, 108
Musculoskeletal disorders, 168

N
National Institute for Occupational Safety and Health (NIOSH), 170
National Institute of Standard and Technology (NIST), 13
National Science Foundation (NSF), 170
Natural user interfaces (NUIs), 123

O
Occupational Safety and Health Administration (OSHA), 135
Off-site automated prefabrication systems, 62, 64, 67
Off-site construction, 18, 78
Off-site fabrication, 77
On-site automated and robotics systems, 62, 64
Operant conditioning (OC), 127

P
Partial Least Squares Structural Equation Modelling (PLS-SEM), 99, 109
Percent score (PS), 37
Physiological signals, 183–185, 188, 189, 191, 196
Physiological status monitoring (PSM), 14
Post-Keynesian theory, 108
Prefabrication, 12, 90
Productivity, 3, 61, 68, 74, 80, 84, 106, 107
Professional responsibility, 106
Public policy, 106

Q
Quality communication, 82
Qualtrics survey software, 38

R
RAC application areas, 62
RAC implementation
  barriers and challenges, 62, 63, 65, 67
  civil infrastructure projects, 65
  construction productivity, 60
  Cronbach's alpha, 63
  evaluation categories, 64
  limiting factors, 68
Radio-frequency (RF), 14
Radio-frequency data communication (RFDC), 14
Radio-frequency identification (RFID), 14
Random forest, 207
Ratings of perceived exertions (RPE), 186
Reality computing technologies
  laser scanners, 8
  rovers, 9
  360-Degree technology, 9
  UASs, 8

Real-time decision-making, 20, 21
Real-time kinematic (RTK), 115
Recycling, 84
Red-Green-Blue-Depth (RGB-D), 119
Research by the National Institute of Occupational Safety and Health, 157
Reusing, 84
Road infrastructure, 98
Robot, 6
Robotic operating system (ROS), 189
Robotic systems, 11, 17
Robotics, 6
Robotics and Automation in Construction (RAC)
  advantages, 60
  challenges, 60
  civil infrastructure (*see* Civil infrastructure projects)
  civil infrastructure projects, 60
  construction professionals, 60
  groups, 61
  on-site adoption, 60
Rovers, 9

S
Safety, 74, 84, 85, 92
Safety impact of drones
  attentional cost
    auditory distractions, 155, 156
    cognitive distractions, 156
    construction environment, 154
    human brains, 154
    type of error, 154
    visual distractions, 155
  equipment/structure, 151
  physical risks
    caught-in/caught-between accidents, 153
    construction sector, 152
    construction sites, 152
    dust and particulates, 154
    indirect/secondary accidents, 154
    number of sources, 152
    struck by accidents, 153
    type of human error, 152
  psychological impact, 157
Safety performance, 166
Sandvik AutoMine products, 12
Science, technology, engineering, and mathematics (STEM), 33
Self-judgment, 80
Semi-automated mason (SAM), 11, 182
Semi-structured interviews, 35, 55
Sensor-based technologies, 20
Single-task construction robots (STCRs), 11, 182
Skyrocketing, 98
Smart construction technology, 12
Social contributions, construction automation
  high-quality living environment, 86, 87
  improved occupational safety and health, 85
  improved working environment, 85
  job satisfaction, 86
  triple bottom lines, 85
Standard deviation (SD), 36
STCR systems approach, 17
Supervised learning algorithms, 207–209
Support vector machine (SVM), 207
Sustainability and construction automation
  economic contributions, 77–82
  environmental contributions, 82–84
  social contributions, 85–87
  substantial environmental concerns, 76
  sustainable development, 76

T
Tachymetry-based methods, 115
Talent cultivation, 32, 43, 51
Technological advancement, 109
Technological aversion, 98
Technology/automation system, 106
Technology competency, 42
Traditional construction, 4
Transition, 105, 108

U
Uniformity, 86
Unmanned aerial systems (UASs), 8, 12, 15, 17, 144
Unmanned ground vehicle (UGV), 184
Urbanization, 2
US Bureau of Labor Statistics (BLS), 167

V
VDC professionals, 15
Virtual reality (VR), 9, 10, 117
Visual line of sight (VLOS), 134
Visualization competency, 42
Visualization technologies
  AR, 10
  DT, 10, 11
  MR, 10
  VR, 9, 10
Volumetric construction, 12
VR-related technologies, 10

W

Waste minimization, 82, 84
Wearable robotics
automation, 166
construction, 166, 168, 169
emerging and innovative intervention opportunity, 167
existing exoskeleton applications, 172, 173
exoskeletons, 175
level of respondents' familiarity, 174
online survey respondent characteristics, 173
research methodology, 171
technology, 166
usability assessment, 175, 176
WMSD risk factors, 167, 168, 174, 175
Wearable sensors, 14
Wireless technology, 149
Work-related musculoskeletal disorders (WMSDs), 166–168

The manufacturer's authorised representative in the EU is Springer Nature Customer Service Centre GmbH, Europaplatz 3, 69115 Heidelberg, Germany. If you have any concerns regarding our products, please contact ProductSafety@springernature.com

Printed and bound by CPI Group (UK) Ltd, Croydon, CR0 4YY
30/09/2026
02225293-0003